# THE NEW GLOBAL ECOSYSTEM IN ADVANCED COMPUTING

Implications for U.S. Competitiveness and National Security

Committee on Global Approaches to Advanced Computing

Board on Global Science and Technology

Policy and Global Affairs Division

**NATIONAL RESEARCH COUNCIL**
*OF THE NATIONAL ACADEMIES*

THE NATIONAL ACADEMIES PRESS
Washington, D.C.
**www.nap.edu**

THE NATIONAL ACADEMIES PRESS   500 Fifth Street, NW   Washington, DC 20001

NOTICE: The project that is the subject of this report was approved by the Governing Board of the National Research Council, whose members are drawn from the councils of the National Academy of Sciences, the National Academy of Engineering, and the Institute of Medicine. The members of the committee responsible for the report were chosen for their special competences and with regard for appropriate balance.

This study was supported by Contract No. HHM402-10-D-0036 between the National Academy of Sciences and the Department of Defense. Any opinions, findings, conclusions, or recommendations expressed in this publication are those of the authors and do not necessarily reflect the views of the organizations or agencies that provided support for the project.

International Standard Book Number-13:  978-0-309-26235-4
International Standard Book Number-10:  0-309-26235-6

Additional copies of this report are available from the National Academies Press, 500 Fifth Street, NW, Keck 360, Washington, DC 20001; (800) 624-6242 or (202) 334-3313; http://www.nap.edu.

Copyright 2012 by the National Academy of Sciences. All rights reserved.

Printed in the United States of America

# THE NATIONAL ACADEMIES
*Advisers to the Nation on Science, Engineering, and Medicine*

The **National Academy of Sciences** is a private, nonprofit, self-perpetuating society of distinguished scholars engaged in scientific and engineering research, dedicated to the furtherance of science and technology and to their use for the general welfare. Upon the authority of the charter granted to it by the Congress in 1863, the Academy has a mandate that requires it to advise the federal government on scientific and technical matters. Dr. Ralph J. Cicerone is president of the National Academy of Sciences.

The **National Academy of Engineering** was established in 1964, under the charter of the National Academy of Sciences, as a parallel organization of outstanding engineers. It is autonomous in its administration and in the selection of its members, sharing with the National Academy of Sciences the responsibility for advising the federal government. The National Academy of Engineering also sponsors engineering programs aimed at meeting national needs, encourages education and research, and recognizes the superior achievements of engineers. Dr. Charles M. Vest is president of the National Academy of Engineering.

The **Institute of Medicine** was established in 1970 by the National Academy of Sciences to secure the services of eminent members of appropriate professions in the examination of policy matters pertaining to the health of the public. The Institute acts under the responsibility given to the National Academy of Sciences by its congressional charter to be an adviser to the federal government and, upon its own initiative, to identify issues of medical care, research, and education. Dr. Harvey V. Fineberg is president of the Institute of Medicine.

The **National Research Council** was organized by the National Academy of Sciences in 1916 to associate the broad community of science and technology with the Academy's purposes of furthering knowledge and advising the federal government. Functioning in accordance with general policies determined by the Academy, the Council has become the principal operating agency of both the National Academy of Sciences and the National Academy of Engineering in providing services to the government, the public, and the scientific and engineering communities. The Council is administered jointly by both Academies and the Institute of Medicine. Dr. Ralph J. Cicerone and Dr. Charles M. Vest are chair and vice chair, respectively, of the National Research Council.

**www.national-academies.org**

**COMMITTEE ON GLOBAL APPROACHES TO ADVANCED COMPUTING**

DANIEL A. REED, Chair, University of Iowa

CONG CAO, University of Nottingham

TAI MING CHEUNG, University of California, San Diego

JOHN CRAWFORD, Intel Corporation

DIETER ERNST, East-West Center

MARK D. HILL, University of Wisconsin–Madison

STEPHEN W. KECKLER, NVIDIA (on sabbatical from the University of Texas at Austin)

DAVID LIDDLE, U.S. Venture Partners

KATHRYN S. MCKINLEY, Microsoft Corporation (on sabbatical from the University of Texas at Austin)


**PRINCIPAL PROJECT STAFF**

WILLIAM O. BERRY, Study Director, Board on Global Science and Technology

ETHAN N. CHIANG, Program Officer, Board on Global Science and Technology

LYNETTE I. MILLETT, Associate Director, Computer Science and Telecommunications Board

**BOARD ON GLOBAL SCIENCE AND TECHNOLOGY**

RUTH DAVID, Chair, Analytic Services, Inc.
HAMIDEH AFSARMANESH, University of Amsterdam
KATY BÖRNER, Indiana University Bloomington
JEFFREY BRADSHAW, Florida Institute for Human and Machine Cognition
DIANNE CHONG, The Boeing Company
JARED COHON, Carnegie Mellon University
ERIC HASELTINE, Haseltine Partners, LLC
JOHN HENNESSY, Stanford University
NAN JOKERST, Duke University
PETER KOLCHINSKY, RA Capital Management, LLC
CHEN-CHING LIU, Washington State University
KIN MUN LYE, Singapore's Agency for Science, Technology and Research
BERNARD MEYERSON, IBM Corporation
KENNETH OYE, Massachusetts Institute of Technology
NEELA PATEL, Abbott Laboratories
DANIEL REED, University of Iowa
DAVID REJESKI, Woodrow Wilson International Center for Scholars

**STAFF**

WILLIAM O. BERRY, Director
PATRICIA WRIGHTSON, Associate Director
ETHAN N. CHIANG, Program Officer
NEERAJ GORKHALY, Research Associate
PETER HUNSBERGER, Financial Officer

# Preface

The information revolution of the last half-century has been driven by dramatic improvements in computing technology—in particular by year-over-year exponential growth in single-processor computing performance that translated into phenomenal new technologies and indeed served as the foundation for entire new industries. Improvements in hardware and associated software advances sustained this growth for decades. In the last few years, those single-processor performance gains have slowed dramatically due to fundamental physical and technical constraints related to power dissipation.[1] Moreover, there is substantial uncertainty as to which technological breakthroughs, if any, may make it possible to continue this approach. This technology disruption has implications not just for the information technology (IT) industry and sectors that depend on it, but for U.S. competitiveness and national security.

The United States has traditionally been on the leading edge of research related to general-purpose computing performance, demonstrated in part by its dominant position in commodity microprocessors for personal computers and servers. The United States has also long been the leader in high-performance computing (HPC) systems, both in research and in deployment. Finally, the United States has also been a leader in the development of graphics processing units (GPUs) and other specialized processors.[2] However, the shift to mobile-based devices and the globalization of the international economy, of communications, and of science and technology (S&T) threatens to erode U.S. technological leadership in these critical areas.

The emergence of global competitors to the United States in advanced computing underscores the need for U.S. policymakers to both understand the advancement of global S&T related to advanced computing and to integrate this understanding with programmatic S&T decision making. To understand these issues more fully, the Office of the Assistant

---

[1] National Research Council, 2011, *The Future of Computing Performance: Game Over or Next Level?*, Washington, D.C.: The National Academies Press (available online at www.nap.edu/catalog.php?record_id =12980). "Before 2004, processor performance was growing by a factor of about 100 per decade; since 2004, processor performance has been growing and is forecasted to grow by a factor of only about 2 per decade. An expectation gap is apparent."

[2] Although both HPC systems and specialized processors are key elements of U.S. competitiveness and national security, the committee's guidance from the sponsor was to focus on the broader computing environment, not on high-end computing. The enabling technologies for these HPC systems are based on the same single-processor, multicore and GPU technologies that are the basis for consumer commodity computing.

Secretary of Defense for Research and Engineering at the Department of Defense asked the National Research Council (NRC) to assess the global S&T landscape for responding to the challenge of improving computing performance in an era where parallel rather than sequential computing is at the forefront.

The Committee on Global Approaches to Advanced Computing was appointed under the auspices of the NRC's Board on Global Science and Technology to conduct this exploration. The nine members of the study committee represent academia and private industry and have expertise in computer science, international S&T, technology assessment, and global innovation. Biographical information for members of the committee is presented in Appendix A. Box P-1 contains committee's statement of task. The committee held three meetings during the course of its work (August, September, and October 2011).

To meet its charge, the committee took a two-part approach. First, it investigated worldwide global research capabilities and commercial competitiveness related to advanced computing,[3] beginning with technology context setting and definitions. As an additional data-gathering experiment, the committee solicited insights from approximately one dozen leading computer scientists and engineers to help identify potential "hubs" of science and technology, relevant to the computing performance challenge (see Appendix B). The committee then examined different innovation strategies, policy tools, and institutional arrangements in a variety of countries that are potentially important players in the development of computing devices technologies and products. Finally, the committee explored the implications of changes in the global advanced computing landscape for U.S. national security.

The data analyses presented in this report are intended to be a starting point for further exploration. The committee's observations highlight important global trends with regard to computing and potential implications for U.S. leadership and for U.S. defense and national security. Rather than providing formal recommendations, this report offers an assessment of the landscape based on the observations and insights of the study committee.

I would like to thank the members of the study committee for their efforts and contributions in developing this report. I also thank the briefers who came and spoke to the committee and provided crucial input and insights that helped to guide our thinking. (Briefers to the committee are listed in Appendix C.) I also thank the reviewers (see Acknowledgment of Reviewers on page xi).

Lastly, the support of the NRC staff was indispensible to accomplishing this study. Special thanks go to Ethan Chiang, who worked closely with the committee throughout the study and played a major role in the preparation of this report. Thanks also go to Lynette Millett for her many valuable insights and contributions.

Daniel Reed,
Chair, Committee on Global Approaches to Advanced Computing

---

[3] By "advanced computing" the committee means any innovations in semiconductor technologies (including fabrication, processing and manufacturing); computer architectures, computing hardware, algorithms and programming approaches; and software developments that improve computing performance or provide new or improved functionality.

**BOX P-1**

**Statement of Task**

An ad hoc committee of the Board on Global Science and Technology (BGST) will describe and assess the global S&T landscape for responding to the challenges of sustaining historical trends in computing performance improvement in general and to the challenge presented by the shift to multicore processors in particular. The committee will identify cutting-edge approaches in computer hardware (e.g., multicore architectures) and software (e.g., emerging parallel programming models) technologies to meet this challenge. The committee will also identify hot spots of innovation around the world and project areas of technological leadership in the United States and elsewhere. Lastly, the committee will consider the implications of these global advances for the U.S. S&T enterprise and for U.S. national security. Based on their work, the committee may suggest criteria or methodologies to more effectively assess the global state of play in a variety of emergent areas of S&T.

To accomplish this task, the committee should consider, but is not limited to, the following questions:

1. What is the cutting edge of approaches for responding to the computing performance challenge?
2. How do other nations and institutions view the computing performance challenge, and what strategies do they have for responding to it?
3. Where are the innovation hot spots in efforts to advance computing performance in the United States and overseas?
4. How are efforts to improve computing performance likely to advance (or stall) over time? Can such efforts be regionally identified? If so, what are they?
5. What are U.S. strengths relative to other international technology leaders in advanced computing performance currently and how might those strengths be expected to change over time?
6. What are the implications of these global advances for U.S. national security in the near and far terms? What are potential resulting IT capabilities and what implications do these have for U.S. national security in the near and far terms?

# Acknowledgment of Reviewers

This report has been reviewed in draft form by individuals chosen for their diverse perspectives and technical expertise, in accordance with procedures approved by the National Research Council's Report Review Committee. The purpose of this independent review is to provide candid and critical comments that will assist the institution in making its published report as sound as possible and to ensure that the report meets institutional standards for objectivity, evidence, and responsiveness to the study charge. The review comments and draft manuscript remain confidential to protect the integrity of the deliberative process. We wish to thank the following individuals for their review of this report:

Eric Archambault, Science-Metrix Inc.;
Mark Bohr, Intel Corporation;
Katy Börner, Indiana University;
Keith Cooper, Rice University;
Peter Cowhey, University of California, San Diego;
Robert Doering, Texas Instruments Incorporated;
Daniel Edelstein, IBM Thomas J. Watson Research Center;
David Kirk, NVIDIA Corporation;
James Larus, Microsoft Research;
David Messerschmitt, University of California, Berkeley;
Henk Moed, Elsevier; and
James Valdes, United States Department of the Army.

Although the reviewers listed above have provided many constructive comments and suggestions, they were not asked to endorse the conclusions or recommendations, nor did they see the final draft of the report before its release. The review of this report was overseen by Anita Jones, University of Virginia and Samuel Fuller, Analog Devices, Inc. Appointed by the National Research Council, they were responsible for making certain that an independent examination of this report was carried out in accordance with institutional procedures and that all review comments were carefully considered. Responsibility for the final content of this report rests entirely with the authoring committee and the institution.

# Contents

SUMMARY     1

1   COMPUTER AND SEMICONDUCTOR TECHNOLOGY TRENDS AND IMPLICATIONS    5
    1.1  Interrelated Challenges to Continued Performance Scaling, 5
    1.2  Future Directions for Hardware and Software Innovation, 11
    1.3  The Rise of Mobile Computing, Services, and Software, 13
    1.4  Summary and Implications, 14

2   THE GLOBAL RESEARCH LANDSCAPE    17
    2.1  Preliminary Observations from Pilot Study of Papers at Top Technical Conferences, 17
    2.2  Increased International Collaboration, 19
    2.3  Commercialization of Technologies, 19
    2.4  Growing Complexity in IT Trade – Tracing Shifts in International Competitiveness, 24
    2.5  China's Position in the Global Semiconductor Value Chain, 28
    2.6  Concluding Remarks, 30

3   INNOVATION POLICY LANDSCAPE – COMPARATIVE ANALYSIS    31
    3.1  Development of the U.S. Computer and Semiconductor Industry, 32
    3.2  China – Strengthening Indigenous Innovation, 36
    3.3  Taiwan – Low-cost and Fast Innovation, 39
    3.4  Korea – Coevolution of International and Domestic Knowledge Linkages, 43
    3.5  Europe – Integrated EU-wide Innovation Policy Coordination, 44
    3.6  Conclusions and Policy Implications, 46

4   IMPLICATIONS OF CHANGES IN THE GLOBAL ADVANCED COMPUTING LANDSCAPE FOR U.S. NATIONAL SECURITY    49
    4.1  Parallelism in Hardware and Software, 49
    4.2  Integrity and Reliability of the Global Supply Chain, 50
    4.3  Decline of Custom Production, 51
    4.4  Convergence of Civilian and Defense Technological Capabilities, 51
    4.5  Rise of a New Post-PC Paradigm Driven by Mass ICT Consumerization, 52
    4.6  New Market-Driven Innovation Centers, 53
    4.7  The Future Educational and Research Landscape in Advanced Computing, 53
    4.8  Cybersecurity and Software, 53
    4.9  Possible Defense IT Outcomes, 54

APPENDIXES 55

    A. Committee Member Biographies 57
    B. Identifying Hubs of Research Activity in Key Areas of S&T Critical to this Study 61
    C. Contributors to the Study 65
    D. Findings and Recommendations from *The Future of Computing Performance: Game Over or Next Level?* 67
    E. Dennard Scaling and Implications 69
    F. Pilot Study of Papers at Top Technical Conferences in Advanced Computing 71
    G. Conference Bibliometric Data 89
    H. Top 20 Largest Hardware and Software Companies 97
    I. China's Medium- and Long-Term Plan 99
    J. List of Abbreviations 101

# Summary

Computing and information and communications technology (ICT) has dramatically changed how we work and live, has had profound effects on nearly every sector of society, has transformed whole industries, and is a key component of U.S. global leadership. A fundamental driver of advances in computing and ICT has been the fact that the single-processor performance has, until recently, been steadily and dramatically increasing year over year, based on a combination of architectural techniques, semiconductor advances, and software improvements. Users, developers, and innovators were able to depend on those increases, translating that performance into numerous technological innovations and creating successive generations of ever more rich and diverse products, software services, and applications that had profound effects across all sectors of society.[1] However, we can no longer depend on those extraordinary advances in single-processor performance continuing.

This slowdown in the growth of single-processor computing performance has its roots in fundamental physics and engineering constraints—multiple technological barriers have converged to pose deep research challenges, and the consequences of this shift are deep and profound for computing and for the sectors of the economy that depend on and assume, implicitly or explicitly, ever-increasing performance. From a technology standpoint, these challenges have led to heterogeneous multicore chips and a shift to alternate innovation axes that include, but are not limited to, improving chip performance, mobile devices, and cloud services. As these technical shifts reshape the computing industry, with global consequences, the United States must be prepared to exploit new opportunities and to deal with technical challenges. The following sections outline the technical challenges, describe the global research landscape, and explore implications for competition and national security.

## Sequential Past, Parallel Future

For multiple decades, single-processor performance has increased exponentially, driven by higher clock rates, reductions in transistor size, faster switching via fabrication improvements, and architectural and software innovations that increased performance while preserving software compatibility with previous-generation processors. This practical manifestation of Moore's Law—the doubling of the number of transistors on a given amount of chip area every 18 to 24 months—created a virtuous cycle of ever-improving single-processor performance and enhanced software functionality.

Hardware and software capabilities and sophistication grew exponentially in part because hardware designers and software developers could innovate in isolation from each other, while still leveraging each other's advances. Software developers created new and more feature-filled applications, confident that new hardware would deliver the requisite performance to execute those applications. In turn, chip designers delivered ever-higher performance chips, while maintaining compatibility with previous generations.

Users benefitted from this hardware-software interdependence in two ways. Not only would old

---

[1] National Research Council, *The Future of Computing: Game Over or Next Level?*, Washington, D.C.: The National Academies Press (available online at http://www.nap.edu/catalog.php?record_id=12980) and NRC, 2003, *Innovation in Information Technology*, Washington, D.C.: The National Academies Press (available online at http://books.nap.edu/catalog.php?record_id=10795).

software execute faster on new hardware, without change, but also new applications exploited advances in graphics and rendering, digital signal processing and audio, networking and communications, cryptography and security—all made possible by hardware advances. Unfortunately, single-processor performance is now increasing at much lower rates—a situation that is not expected to change in the foreseeable future.

The causes for the declining rates of chip hardware performance improvements begin with the limit on chip power consumption, which is proportional to the product of the chip clock frequency and the square of the chip operating voltage. As chip clock frequencies rose from megahertz to gigahertz, chip vendors improved fabrication processes and reduced chip operating voltages and, thus, power consumption.

However, it is no longer practical to increase performance via higher clock rates, due to power and heat dissipation constraints. These constraints are themselves manifestations of more fundamental challenges in materials science and semiconductor physics at increasingly small feature sizes. While the market for the highest performance server processor chips continues to grow, the market demand for phones, tablets, and netbooks has also increased emphasis on low-power, energy-efficient processors that maximize battery lifetime.

Finally, the use of additional transistors to preserve the sequential instruction execution model while accelerating instruction execution reached the point of diminishing returns. Indeed, most of the architectural ideas that were once found only in exotic supercomputers (e.g., deep pipelines, multiple instruction issue, out-of-order instruction logic, branch prediction, data and instruction prefetching) are commonplace within microprocessors.

The combination of these challenges—power limitations, diminishing architecture returns, and semiconductor physics challenges—drove a shift to multicore processors (i.e., placing multiple processors, sometimes of differing power or performance and function, on a single chip). By making parallelism visible to the software, this technological shift disrupted the cycle of sequential performance improvements and software evolution atop a standard hardware base.

Beginning with homogeneous multicore chips (i.e., multiple copies of the same processor core), design alternatives are evolving rapidly, driven by the twin constraints of energy efficiency and high performance. In addition, system-on-a-chip designs are combining heterogeneous hardware functions used in smartphones, tablets, and other devices. The result is a dizzying variety of parallel functionality on each chip. It is likely that even more heterogeneity will arise from expanded use of accelerators and reconfigurable logic for increased performance while simultaneously meeting power constraints.

Whether homogeneous or heterogeneous, these chips are dependent on parallel software for operation, for there is no known alternative to parallel programming for sustaining growth in computing performance. However, unlike in the sequential case, there is no universally accepted, compelling programming paradigm for parallel computing. Absent such programming models and tools, creating increasingly sophisticated applications that fully and effectively exploit parallel chips is difficult at best. Thus, there exists a great opportunity and need for renewed research on parallel algorithms and programming methodologies, recognizing that this is a challenge and long-studied problem. However, because multicore chips are dependent on parallel programming, it is prudent to continue such explorations.

Although further research in parallel programming models and tools may ameliorate this problem (e.g., via domain-specific languages, high-level libraries, and toolkits), 40 years of research in parallel computing suggests this outcome is by no means certain. When combined with the need for increasingly rapid development cycles to respond to changing demands and the rising importance of software security and resilience in an Internet-connected world, the programming challenges are daunting. In combination, the continued slowing of processor performance and the uncertainty of a parallel software future poses potential short- and long-term risks for U.S. national security and the U.S. economy. This report focuses on the competitive position of the U.S. semiconductor and software industries and their impact on U.S. national security in the new norm of parallel computing.

## Global Competition and the Research Landscape

Because of this disruption to the computing ecosystem,[2] major innovations in semiconductor processes, computer architecture, and parallel programming tools and techniques are all needed if we are to continue to deliver ever-greater application performance.

---

[2] The advanced computing ecosystem refers not only to the benefits from and interdependencies between breakthroughs in academic and industry science and engineering research and commercialization success by national, multi-national and global companies, but also the underlying infrastructure (that includes components such as workforce; innovation models, e.g., centralist versus entrepreneurial; global knowledge networks; government leadership and investment; the interconnectedness of economies; and global markets) that underpin technological success.

In the past, the U.S. Department of Defense's (DOD) uptake of U.S. computing technology research designed especially for it and now increasingly adapted from the fast-moving consumer market has resulted in a large U.S. advantage. In the future, the rate of change in the competitive position of the United States in computing technology will increasingly depend in part on other countries' basic research capabilities and the types of research and development (R&D) policies they pursue, as well as the associated economic climate. Of course, many factors influence the range and type of policy options available in each region. Countries also differ in their levels of development and in their economic institutions, and pursue quite different approaches to innovation policy.

Historically, the United States has relied on market forces and the private sector to convert university research ideas, funded by the federal government, into marketable products. In contrast, the European Union and emerging economies such as China, Korea, and Taiwan rely much more on the government to define the strategic objectives and key parameters. For example, recent Chinese innovation policies have played an increasing role in strengthening its indigenous innovation capabilities. There is also evidence that China is transitioning toward economic outcome-driven science and technology programs focused on technologies of national strategic importance—many of which are advanced computing technologies. In contrast, Taiwan's innovation policies are focused on moving its IT industries beyond the traditional "global factory" model. Thus, innovation polices emphasize low-cost and fast innovation by strengthening public and private partnerships that leverage domestic and global innovation networks.

**Competitive Implications and National Security**

In the committee's view, the United States currently enjoys a technological advantage in many computing technologies. Nonetheless, this technological gap is narrowing as other countries, such as China, make a concerted effort to develop their own indigenous computing design and manufacturing capabilities and as design and fabrication of such technologies, as well as software development, are increasingly distributed globally.

Thus, it is important to take a long-term perspective on our approaches to computing innovation, technology uptake, and defense policy, for the United State's global competitors certainly are. The principal future national security concerns for the United States related to anticipated computing shifts and limits on single-processor performance come not just from the threat to U.S. technological superiority, but also from changes to the nature and structure of the marketplace for computing and information technology. U.S. challenges include maintaining the integrity of the global supply chain for semiconductors, which is exacerbated by the convergence of civilian and defense technologies, as well as the rise of a new ecosystem of smart devices, based on licensable components and created by semiconductor design firms without fabrication capabilities.

Over time, the increasing presence and establishment of foreign markets that are larger, are potentially more lucrative, and have better long-term growth potential than in the United States and other developed countries could also have significant implications. Any shift in the global commercial center of gravity could lead to a shift in the global R&D center of gravity as international firms are required to locate in these markets if they are to remain competitive and to meet the requirements of government regulations in the target markets.

Shifting from policy to technology, the parallel programming challenges in delivering high performance on multicore chips are real and global, with no obvious technical solutions. Barring research breakthroughs, developing applications that exploit on-chip parallelism effectively (or vice versa, by developing approaches to on-chip parallelism that better support application needs) will remain an intellectually challenging task that is dependent on highly skilled software developers. When combined with the need for rapid application development, nimble response to shifting threats, and the ever-present desire for new features, equating competitive advantage in computing solely with single-processor performance (and associated application performance) may not be wise. Going forward, metrics such as system reliability, energy efficiency, security adaptability, and cost will inevitably become more salient. Power consumption is the major constraint on chip performance and device utility. Innovation in software, architecture, hardware, and other computing technologies will continue apace, but the primary axes of innovation are shifting, and organizations such as the U.S. DOD will need to adapt their computing and IT strategies accordingly.

# 1

# Computer and Semiconductor Technology Trends and Implications

Computing and information and communications technology has had incredible effects on nearly every sector of society. Until recently, advances in information and communications technology have been driven by steady and dramatic gains in single-processor (core) speeds. However, current and future generations of users, developers, and innovators will be unable to depend on these improvements in computing performance.

In the several decades leading up to the early 2000s, single-core processor performance doubled about every 2 years. These repeated performance doublings came to be referred to in the popular press as "Moore's Law," even though Moore's Law itself was a narrow observation about the economics of chip lithography feature sizes.[1] This popular understanding of Moore's Law was enabled by both technology—higher clock rates, reductions in transistor size, and faster switching via fabrication improvements—and architectural and compiler innovations that increased performance while preserving software compatibility with previous-generation processors. Ongoing and predictable improvements in processor performance created a cycle of improved single-processor performance followed by enhanced software functionality. However, it is no longer possible to increase performance via higher clock rates, because of power and heat dissipation constraints. These constraints are themselves manifestations of more fundamental challenges in materials science and semiconductor physics at increasingly small feature sizes.

A National Research Council (NRC) report, *The Future of Computing Performance: Game Over or Next Level?*,[2] explored the causes and implications of the slowdown in the historically dramatic exponential growth in computing performance and the end of the dominance of the single microprocessor in computing. The findings and recommendations from that report are provided in Appendix D. The authoring committee of this report concurs with those findings and recommendations. This chapter draws on material in that report and the committee's own expertise and discusses the technological challenges to sustaining growth in computing performance and their implications for computing and innovation. The chapter concludes with a discussion of the implications of these technological realities for United States defense. Subsequent chapters have a broader emphasis, beyond technology, on the implications for global technology policy and innovation issues.

## 1.1 Interrelated Challenges to Continued Performance Scaling

The reasons for the slowdown in the traditional exponential growth in computing performance are many. Several technical drivers have led to a shift from ever-faster single-processor computer chips as the foundation for nearly all computing devices to an emphasis on what have been called "multicore" processors—placing

---

[1] The technological and economic challenges are intertwined. For example, Moore's Law is enabled by the revenues needed to fund the research and development necessary to advance the technology. See, for example, *The Economic Limit to Moore's Law – IEEE Transactions on Semiconductor Manufacturing*, Vol. 24, No. 1, February 2011.

[2] NRC, *The Future of Computing Performance: Game Over or Next Level?*, Washington, D.C.: The National Academies Press (available online at http://www.nap.edu/catalog.php?record_id=12980.

multiple processors, sometimes of differing power and/or performance characteristics and functions, on a single chip. This section describes those intertwined technical drivers and the resulting challenges to continued growth in computing performance. This shift away from an emphasis on ever-increasing speed has disrupted what has historically been a continuing progression of dramatic sequential performance improvements and associated software innovation and evolution atop a predictable hardware base followed by increased demand for ever more software innovations that in turn motivated hardware improvements. This disruption has profound implications not just for the information technology industry, but for society as a whole. This section first describes the benefits of this virtuous cycle—now ending—that we have depended on for so long. The technical challenges related to scaling nanometer devices, what the shift to multicore architectures means for architectural innovation, programming explicitly parallel hardware, increased heterogeneity in hardware, and the need for correct, secure, and evolvable software are then discussed.

### 1.1.1 Hardware-Software Virtuous Cycle

The hardware and performance improvements described above came with a stable programming interface between hardware and software. This interface persisted over multiple hardware generations and in turn contributed to the creation of a virtuous hardware-software cycle (see Figure 1-1). Hardware and software capabilities and sophistication each grew dramatically in part because hardware and software designers could innovate in isolation from each other, while still leveraging each other's advances in a predictable and sustained fashion. For example, hardware designers added sophisticated out-of-order instruction issue logic, branch prediction, data prefetching, and instruction prefetching to the capabilities. Yet, even as the hardware became more complex, application software did not have to change to take advantage of the greater performance in the underlying hardware and, consequently, achieve greater performance on the software side as well.

Software designers were able to make grounded and generally accurate assumptions about future capabilities of the hardware and could—and did—create software that needed faster, next-generation processors with larger memories even before chip and system architects actually were able to deliver them. Moreover, rising hardware performance allowed software tool developers to raise the level of abstraction for software development via advanced libraries and programming models, further accelerating application development. New, more demanding applications that only executed on the latest, highest performance hardware drove the market for the newest, fastest, and largest memory machines as they appeared.

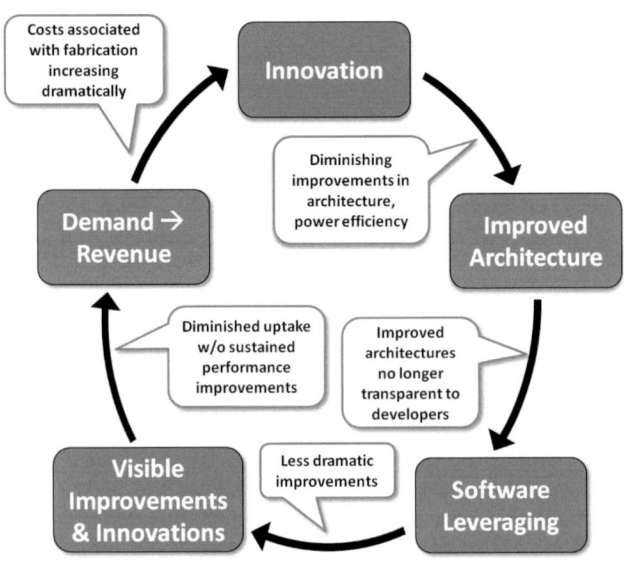

FIGURE 1-1 Cracks in the hardware-software virtuous cycle. SOURCE: Adapted from a 2011 briefing presentation on the Computer Science and Telecommunications Board report *The Future of Computing Performance: Game Over or Next Level?*

Another manifestation of the virtuous cycle in software was the adoption of high-level programming language abstractions, such as object orientation, managed runtimes, automatic memory management, libraries, and domain-specific languages. Programmers embraced these abstractions (1) to manage software size, sophistication, and complexity and (2) to leverage existing components developed by others. However, these abstractions are not without cost and rely on system software (i.e., compilers, runtimes, virtual machines, and operating systems) to manage software complexity and to map abstractions to efficient hardware implementations. In the past, as long as the software used a sequential programming interface, the cost of abstraction was hidden by ongoing, significant improvements in hardware performance. Programmers embraced abstraction and consequently produced working software faster.

Looking ahead, it seems likely that the right choice of new abstractions will expand the pool of programmers further. For example, a domain specialist can become a programmer if the language is intuitive and the abstractions match his or her domain expertise well. Higher-level abstractions and domain-specific toolkits, whether for technical computing or World Wide Web services, have allowed software developers to create complex systems quickly and with fewer common errors. However, implicit in this approach has been an assumption that hardware performance would continue to increase (hiding the overhead of these abstractions) and that developers need not understand the mapping of the abstractions to hardware to achieve adequate performance.[3] As these assumptions break down, the difficulty in achieving high performance from software will rise, requiring hardware designers and software developers to work together much more closely and exposing increasing amounts of parallelism to software developers (discussed further below). One possible example of this is the use of computer-aided design tools for hardware-software co-design. Another source of continued improvements in delivered application performance could also come from efficient implementation techniques for high-level programming language abstractions.

*1.1.2 Problems in Scaling Nanometer Devices*

Early in the 2000s, semiconductor scaling—the process of technology improvement so that it performs the same functionalities at ever smaller scales—encountered fundamental physical limits that now make it impractical to continue along the historical paths to ever-increasing performance.[4] Expected improvements in both performance and power achieved with technology scaling have slowed from their historical rates, whereas implicit expectations were that chip speed and performance would continue to increase dramatically. There are deep technical reasons for (1) why the scaling worked so well for so long and (2) why it is no longer delivering dramatic performance improvements. See Appendix E for a brief overview of the relationship between slowing processor performance growth and Dennard scaling and the powerful implications of this slowdown.

In fact, scaling of semiconductor technology hit several coincident roadblocks that led to this slowdown, including architectural design constraints, power limitations, and chip lithography challenges (both the high costs associated with patterning smaller and smaller integrated circuit features and with fundamental device physics). As described below, the combination of these challenges can be viewed as a perfect storm of difficulty for microprocessor performance scaling.

With regard to power, through the 1990s and early 2000s the power needed to deliver performance improvements on the best performing microprocessors grew from about 5–10 watts in 1990 to 100–150 watts in 2004 (see Figure 1-2). This increase in power stopped in 2004, because cooling and heat dissipation proved inadequate. Furthermore, the exploding demand for portable devices, such as phones, tablets, and netbooks, increased the market importance of lower-power and energy-efficient processor designs.

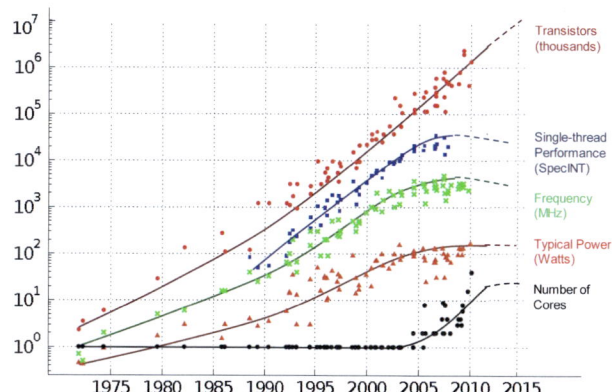

FIGURE 1-2 Thirty five years of microprocessor trend data.
SOURCE: Original data collected and plotted by M. Horowitz, F. Labonte, O. Shacham, K. Olukotun, L. Hammond, and C. Batten. Dotted-line extrapolations by C. Moore: Chuck Moore, 2011, "Data processing in exascale-class computer systems," *The Salishan Conference on High Speed Computing*, April 27, 2011. (www.lanl.gov/orgs/hpc/salishan)

In the past, computer architects increased performance with clever architectural techniques such as ILP (instruction-level parallelism through the use of deep pipelines, multiple instruction issue, and speculation) and memory locality (multiple levels of caches). As the number of transistors per unit area on a chip continued to increase (as predicted by Moore's Law), microprocessor designers used these transistors to, in part, increase the potential to exploit ILP by increasing the number of instructions executed in parallel (IPC, or instructions per

---

[3] Such abstractions may increase the energy costs of computation over time; a focus on energy costs (as opposed to performance) may have led to radically different strategies for both hardware and software. Hence, energy-efficient software abstractions are an important area for future development.

[4] In "High-Performance Processors in a Power-Limited World," Sam Naffziger reviews the $V_{dd}$ limitations and describes various approaches (circuit, architecture) to future processor design given the voltage scaling limitations: Sam Naffziger, 2006, "High-performance processors in a power-limited world," *Proceedings of the IEEE Symposium on VLSI Circuits*, Honolulu, HI, June 15–17, 2006, p. 93–97.

clock cycle).[5] Transistors were also used to achieve higher frequencies than were supported by the raw transistor speedups, for example, by duplicating logic and by reducing the depth of logic between pipeline latches to allow faster clock cycles. Both of these efforts yielded diminishing returns in the mid-2000s. ILP improvements are continuing, but also with diminishing returns.[6]

Continuing the progress of semiconductor scaling—whether used for multiple cores or not—is now dependent on innovation in structures and materials to overcome the reduced performance scaling traditionally provided by Dennard scaling.[7]

Continued scaling also depends on continued innovation in lithography. Current state-of-the-art manufacturing uses a 193-nanometer wavelength to print structures that are only tens of nanometers in size. This apparent violation of optical laws has been supported by innovations in mask patterning and compensated for by increasingly complex computational optics. Future lithography scaling is dependent on continued innovation.

*1.1.3 The Shift to Multicore Architectures and Related Architectural Trends*

The shift to multicore architectures meant that architects began using the still-increasing transistor counts per chip to build multiple cores per chip rather than higher-performance single-core chips. Higher-performance cores were eschewed in part because of diminishing performance returns and emerging chip power constraints that made small performance gains at a cost of larger power use unattractive. When single-core scaling slowed, a shift in emphasis to multicore chips was the obvious choice, in part because it was the only alternative that could be deployed rapidly. Multicore chips consisting of less complex cores that exploited only the most effective ILP ideas were developed. These chips offered the promise of performance scaling linearly with power. However, this scaling was only possible if software could effectively make use of them (a significant challenge). Moreover, early multicore chips with just a few cores could be used effectively at either the operating system level, avoiding the need to change application software, or by a select group of applications retargeted for multicore chips.

With the turn to multicore, at least three other related architectural trends are important to note to understand how computer designers and architects seek to optimize performance—a shift toward increased data parallelism, accelerators and reconfigurable circuit designs, and system-on-a-chip (SoC) integrated designs.

First, a shift toward increased data parallelism is evident particularly in graphics processing units (GPUs). GPUs have evolved, moving from fixed-function pipelines to somewhat configurable ones to a set of throughput-oriented "cores" that allowed more successful general-purpose GPU (GP-GPU) programming.

Second, accelerators and reconfigurable circuit designs have matured to provide an intermediate alternative between software running on fixed hardware, for example, a multicore chip, and a complete hardware solution such as an application-specific integrated circuit, albeit with their own cost and configuration challenges. Accelerators perform fixed functions well, such as encryption-decryption and compression-decompression, but do nothing else. Reconfigurable fabrics, such as field-programmable gate arrays (FPGAs), sacrifice some of the performance and power benefits of fixed-function accelerators but can be retargeted to different needs. Both offer intermediate solutions in at least four ways: time needed to design and test, flexibility, performance, and power.

Reconfigurable accelerators pose some serious challenges in building and configuring applications; tool chain issues need to be addressed before FPGAs can become widely used as cores. To use accelerators and reconfigurable logic effectively, their costs must be overcome when they are not in use. Fortunately, if power, not silicon area, is the primary cost measure,

---

[5] Achieved application performance depends on the characteristics of the application's resource demands and on the hardware.

[6] ILP improvements are incremental (10–20 percent), leading to single-digit compound annual growth rates.

[7] According to Mark Bohr, "Classical MOSFET scaling techniques were followed successfully until around the 90nm generation, when gate-oxide scaling started to slow down due to increased gate leakage" (Mark Bohr, February 9, 2009, "ISSCC Plenary Talk: The New Era of Scaling in an SOC World") At roughly the same time, subthreshold leakage limited the scaling of the transistor $V_t$ (threshold voltage), which in turn limited the scaling of the voltage supply in order to maintain performance. Since the active power of a circuit is proportional to the square of the supply voltage, this reduced scaling of supply voltage had a dramatic impact on power. This interaction between leakage power and active power has led chip designers to a balance where leakage consumes roughly 30 percent of the power budget. Several approaches are being undertaken. *Copper interconnects* have replaced aluminum. *Strained silicon* and *Silicon-on-Insulator* have provided improved transistor performance. Use of a *low-K dielectric material* for the interconnect layers has reduced the parasitic capacitance, improving performance. *High-K metal gate transistor structures* restarted gate "oxide" scaling with orders of magnitude reduction in gate leakage. Transistor structures such as *FinFET*, or Intel's *Tri-Gate* have improved control of the transistor channel, allowing additional scaling of $V_t$ for improved transistor performance and reduced active and leakage power.

turning the units off when they are not needed reduces energy consumption (see discussion of dark and dim silicon, below).

Third, increasing levels of integration that made the microprocessor possible four decades ago now enable complete SoCs. They combine most of the functions of a motherboard onto a single chip, usually with off-chip main memory. These processors integrate memory and input/output controllers, graphics processors, and other special-purpose accelerators. These (SoC) designs are widely used in almost all devices, from servers and personal computers to smartphones and embedded devices.

Fourth, power efficiency is increasingly a major factor in the design of multicore chips. Power has gone from a factor to optimize in the near-final design of computer architectures to a second-order constraint to, now, a first-order design constraint. As the right side of Figure 1-2 projects, future systems cannot achieve more performance from simply a linear increase in core count at a linear increase in power. Chips deployed in everything from phones, tablets, and laptops to servers and data centers must take into account power needs.

One technique for enabling more transistors per chip at better performance levels without dramatically increasing the power needed per chip is dark silicon. Dark silicon refers to a design wherein a chip has many transistors, but only a fraction of them are powered on at any one time to stay within a power budget. Thus, function-specific accelerators can be powered on and off to maximize chip performance. A related design is dim silicon where transistors operate in a low-power but still-useful state. Dark and dim silicon make accelerators and reconfigurable logic more effective. However, making dark and dim silicon practical is not easy, because adding silicon area per chip always raises cost, even if the silicon only provides value when it is on. This also presents significant software challenges, as each heterogeneous functional unit requires efficient code (e.g., this may mean multiversion code, as well as compilers and tool chains designed for many variations). Thus, even as dark and dim silicon become more widely adopted, using them to create value is a significant open challenge.

Moreover, emerging transistors have more variability than in the past, due to variations in the chip fabrication process: Some transistors will be faster, while others are slower, and some use more power and others use less. This variability is emerging now, because some aspects of fabrication technology (e.g., gate oxides) are reaching atomic dimensions. Classically, hardware hid almost all errors from software (except memory errors) with techniques (such as guard bands) that conservatively set parameters well above a mean value to tolerate variation while creating the illusion of error-free hardware. As process variation grows relative to mean values, guard bands become overly conservative. This means that new errors will be exposed more frequently to software, posing software and system reliability challenges.

*1.1.4 Game Changer: Programming for Explicitly Parallel Commodity Hardware*

The advent of multicore chips changes the software interface. Sequential software no longer becomes faster with every hardware generation, and software needs to be written to leverage parallel hardware explicitly. Current trends in hardware, specifically multicore, might seem to suggest that every technology generation will increase the number of processors and, accordingly, that parallel software written for these chips would speed up in proportion to the number of processors (often referred to as scalable software).

Reality is not so straightforward. There are limits to the number of cores that can usefully be placed on a chip. Moreover, even software written in parallel languages typically has a sequential component. In addition, there are intrinsic limits in the theoretically available parallelism in some problems, as well as in their solution via currently known algorithms. Even a small fraction of sequential computation significantly compromises scalability (see Figure 1-3), compromising expected improvements that might be gained by additional processors on the chip.

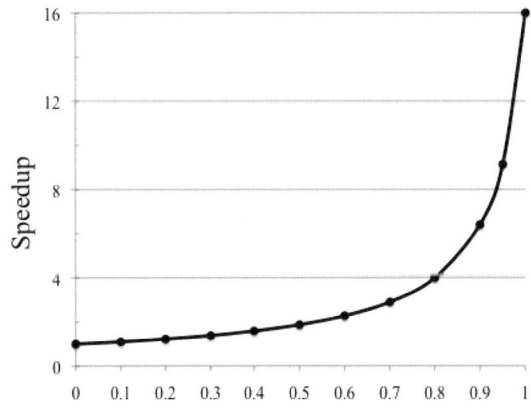

FIGURE 1-3 Amdahl's Law example of potential speedup on 16 cores based on the fraction of the program that is parallel.

As part of ongoing research programs, it will be important to analyze the interplay among the available parallelism in applications, energy consumption by the resulting chip (under load with real applications), the performance of different algorithmic formulations, and programming complexity.

In addition, most programs written in existing parallel languages are dependent on the number of hardware processors. Further developments in parallel software are necessary to be performance portable, that is, it should execute on a variety of parallel computing platforms and should show performance in proportion to the number of processors on all these platforms without modifications, within some reasonable bounds.

Deeply coupled to parallelism is data communication. To operate on the same data in parallel on different processors, the data must be communicated to each processor. More processors imply more communication. Communicating data between processors on the same chip or between chips is costly in power and time. Unfortunately, most parallel programming systems result in programs whose performance heavily depends on the memory hierarchy organization of the processor. Where the data is located in a system directly affects performance and energy. Consequently, sequential and even existing parallel software is not performance portable to successive generations of evolving parallel hardware, or even between two machines of the same generation with the same number of processors if they have different memory organizations. Software designers currently must modify software for it to run efficiently on each multicore machine. The need for such efforts breaks the virtuous cycle described above and makes building and evolving correct, secure, and performance-portable software a substantial challenge.

Finally, automatic parallelization systems, which seek to transform a sequential program into a parallel programmer without programmer intervention, have mostly failed. Had they been successful, programmers would be able to write in a familiar sequential language and yet still see the performance benefits of parallel execution. As a result, research now focuses on programmer-specified parallelism.

### 1.1.5 Heterogeneity in Hardware

As mentioned above, not only are technology trends leading designers and developers of computer hardware to focus on multicore systems, but they are also leading to an emphasis on specialization and heterogeneity to provide power, performance, and energy efficiency. This specialization is a marked contrast to previous approaches. GPUs are an example of hardware specialization designed to be substantially more power efficient for a specific workload. The problem with this trend is three-fold.

First, hardware specialization can only be justified for ubiquitous and/or high-value workloads due to the high cost of chip design and fabrication. Second, creating software that exploits hardware specialization and heterogeneity closely couples hardware and software—such coupling may be good for performance, power, and energy, but it typically sacrifices software portability to different hardware, a mainstay expectation in computing over many decades.

Third, the lead time needed for effective software support of these heterogeneous devices may reduce the time they can be competitive in the marketplace. If it takes longer to deliver the tools (compilers, domain-specific language, and so on) than it takes to design and deliver the chip, then the tools will appear after the chip, with negative consequences.[8] This problem, however, is not new. For example, by holding the IA-32 instruction set architecture relatively constant across generations of hardware, software could be delivered in a timely manner. Designing and building a software system for hardware that does not exist, or is not similar to prior hardware, requires well-specified hardware-software interfaces and accurate simulators to test the software independently. Because executing software on simulators requires tens to thousands of more time than executing on actual hardware, software will lag hardware without careful system and interface design. In summary, writing portable and high-performance software is hard, making such software parallel is harder, and developing software that can exploit heterogeneous parallel architectures is even harder.

### 1.1.6 Correct and Secure Software that Evolves

Performance—in the sense of ever-increasing chip speed—is not the only critical demand of modern application and system software. Although performance is fungible and ever-faster computer chips can be used to enable a variety of functionality, software is the underpinning of virtually all our economic, government, and military infrastructure. Thus, the criticality of *S*ecure, *P*arallel *E*volvable, *R*eliable, and *C*orrect software cannot be overemphasized. This report uses the term *SPERC software* to refer to these software properties.

---

[8]The same is true for hardware-software co-design efforts. Success in co-design requires that both the hardware and software be delivered at roughly the same time. If the software lags behind the hardware, it diminishes the strategy's effectiveness.

Achieving each of these desirable SPERC software properties is difficult in isolation, and each property is still the subject of much research. In an era where new technologies—at all levels of the system—appear quickly, yet the rate of hardware performance improvement is slowing, an alternative to the virtuous cycle described earlier is essential. Rather than remaining oblivious to hardware shifts, new approaches and methodologies are needed that allow our complex software systems to *evolve* nimbly, using new technologies and adapting to changing conditions for rapid deployment. This flexibility and rapid adaptation will be key to continued superiority, for all large-scale enterprises, including military and defense needs.

In addition to flexibility and nimbleness, as the world becomes more connected, building software that executes *reliably* and guarantees some *security* properties is critical. For example, modern programming systems for languages such as PHP, JavaScript, Java, and C#, while more secure than native systems because of their type and memory safety, do not guarantee provably secure programs. For example, mainstream programming models do not yet support concise expression of semantic security properties such as "only an authenticated user can access their own data," which is key to proving security properties. Even recipes of best practices for secure programming remain an open problem.

Finally, functional correctness remains a major challenge. Designing and building *correct* parallel software is a daunting task. For example, static verification is the process that analyzes code to ensure that it guarantees certain properties and user-defined specifications. Static verification of even basic properties of sequential software in some cases cannot be decided, and computing approximations often involves exponential amounts of computation to analyze properties on all programming paths. Evaluating the same properties in *parallel* programs is even harder, since the analysis must consider all possible execution interleaving of concurrent statements in distinct parallel tasks. Current practice sometimes verifies small critical components of large systems, but for the most part, executes the program on a variety of test inputs (testing) to detect errors. Correctness and security demands on software may trump performance in some cases, but applications will typically need to combine these properties with high performance and parallelism.

Even assuming that there are programming models that establish a solid foundation for creating SPERC software, adoption will be a challenge. Commodity and defense software will need to be created or ported to use them. The enormous investment in legacy software and the large cost of porting software to new languages and platforms will be a barrier to adoption.

## 1.2 Future Directions for Hardware and Software Innovation

Section 1.1 outlined many of the technological challenges to continued growth in computing performance and some of the implications (e.g., the shift to multicore and increased emphasis on power efficiency.) This section provides a brief overview of current hardware and software research strategies for building and evolving future computer systems that seek continued improvements to high performance and energy efficiencies.

### *1.2.1 Advanced Hardware Technology Options*

Earlier sections of this chapter described issues that have hindered continued scaling of modern semiconductor technology and some of the current innovations in materials and structures that have allowed continued progress. All are variations on historical approaches. Are there more radical innovations that may deliver future improvements? In principle, yes, but there are daunting challenges.

Transistors built from alternative materials such as germanium (Ge) and Group III–V materials, such as gallium arsenide, indium phosphide, indium arsenide, and indium antimonide, promise improved power efficiency,[9] but only by about a factor of two, as they also suffer from the same threshold voltage limits, and limit on-supply voltage scaling inherent in current complementary-symmetry metal-oxide semiconductor technologies.

Advances in packaging technology continue, and some of those offer promise for power and performance improvements. For example, 3D stacking and through-silicon vias are being explored for some SoC designs. The primary limitation for 3D stacking of memory, however, is capacity (i.e., only limited dynamic random-access memory can be placed in the stack).

Finally, more exotic alternatives to the use of electrons as the "tokens," coupled with an energy barrier as the control—the method used by all modern computer chips—are under investigation.[10] Although all of these

---

[9]Donghyun Kim, Tejas Krishnamohan, and Krishna C. Saraswat, 2008, "Performance Evaluation of 15nm Gate Length Double-Gate n-MOSFETs with High Mobility Channels: III–V, Ge and Si," *The Electroch. Soc. Trans.* 16(11): 47–55.

[10]K. Bernstein, R. Calvin, W. Porod, A. Seabaugh, and J. Welser, 2010, "Device and Architecture Outlook for Beyond CMOS Switches," *Proceedings of the IEEE* 98(12): 2169–2184.

technologies show potential, each has serious challenges that need to be resolved through continued fundamental research before they could be adopted for high-volume manufacturing.

*1.2.2 Prospects for Performance Improvements*

In the committee's view, there is no "silver bullet" to address current challenges in computer architecture and the slowdown in growth in computing performance. Rather, efforts in complementary directions to continue to increase value, that is, performance, under power constraints, will be needed. Early multicore chips offered homogeneous parallelism. Heterogeneous cores on a single chip are now part of an effort to achieve greater power efficiency, but they present even greater programming challenges.

Efforts to advance conventional multicore chips and to create more power-efficient core designs will continue. On one hand, researchers will continue to explore techniques that reduce the power used by individual cores without unduly sacrificing performance. In turn, this will allow placement of more cores on each chip. Researchers could also explore radical redesigns of cores that focus on power first, for example, by minimizing nonlocal information movement through spatially aware designs that limit communication of data (see Section 1.1.4).

GP-GPU computing, in particular, and vector and single-instruction multiple-data operation, in general, offer promise for specific workloads. Each of these reduce power consumption by amortizing the cost of dealing with an instruction (e.g., fetch and decode) across the benefit of many data manipulations. All offer great peak performance, but this performance can be hard to achieve without deep expertise coupling algorithm and architecture, hardly a prescription for broad programmability. Moreover, software that runs on such chips must allocate work to cope with allocating work to heterogeneous computing units, such as throughput-oriented GPUs and latency-oriented conventional central processing units, highlighting the need for advances in software and programming methodologies as described earlier.

More heterogeneity will arise from expanded use of accelerators and reconfigurable logic, described earlier, that is needed for increased performance under power constraints. Accelerators are so-named because they can accelerate performance. While this is true, recent work shows that the greater benefit of accelerators may be in reducing power.[11] However, accelerator effectiveness can be blunted by the overheads of communicating control and data to and from accelerators, especially if someone seeks to offload even smaller amounts of work to expand the availability of off-loadable work. Reconfigurable designs, such as FPGAs, described earlier, may provide a middle ground, but they are not yet easily programmable. Similarly, SoCs combine specialized accelerators on a single chip and have had great success in the embedded market, such as smartphones and tablets. As SoCs continue to proliferate, the challenge will be simplifying software and hardware design and programmability while maximizing performance and power efficiency.

Moreover, communication at all levels—close, cross-chip, off-chip, off-board, off-node, offsite—must be minimized to save energy. For example, moving operands from a close-by register file can use energy comparable to an operation (e.g., floating-point multiply-add), while moving them from cross- or off-chip uses tens to hundreds of times more energy. Thus, a focus on reducing computational energy without a concomitant focus on reducing communication is doomed to have limited effect.

Finally, a reconsideration of the hardware-software boundary may be in order. While abstraction layers hide complexity at each boundary, they also hide optimization and innovation possibilities. For decades, software and hardware experts innovated independently on opposite sides of the instruction set architecture boundary. Multicore chips began the end of the era of separation. Going forward, co-design is needed, where chip functionality and software are designed in concert, with repeated design and optimization feedback between the hardware and software teams. However, since the software development cycle typically significantly lags behind the hardware development cycle, effective co-design will also require more rapid deployment of effective tools in a timescale commensurate with the specialized hardware if its full functionality is to be realized.

*1.2.3 Software*

Creating software systems and applications for parallel, power-constrained computing systems on a single chip requires innovations at all levels of the software-hardware design and engineering stack: algorithms, programming models, compilers, runtime

---

[11] Rehan Hameed, Wajahat Qadeer, Megan Wachs, Omid Azizi, Alex Solomatnikov, Benjamin C. Lee, Stephen Richardson, Christos Kozyrakis, and Mark Horowitz, 2010, "Understanding Sources of Inefficiency in General-Purpose Chips," *Proceedings of the 37th International Symposium on Computer Architecture (ISCA)*, Saint-Malo, France, June 2010.

systems, operating systems, and hardware-software interfaces.

One strategy for addressing the challenges inherent to parallel programming is to first design application-specific languages and system software and then seek generalizations. The most successful examples of parallelism come from distributed search systems, Web services, and databases executing on distinct devices, as opposed to the challenge of parallelism within a single device (chip) that is addressed here. Parallel algorithm and system design success stories include MapReduce[12] for processing data used in search, databases for persistent storage and retrieval, and domain-specific toolkits with parallel support, such as MATLAB. Part of their success is rooted in providing a level of abstraction in which programmers write sequential components, while the runtime and system software implement and manage the parallelism. On the other hand, GPU programming is also a success story, but when used for game engineering development, for instance, it relies on expert programmers with deep knowledge of parallelism, algorithm-to-hardware mappings, and performance tuning. General-purpose computing on GPUs does not require in-depth knowledge about graphics hardware, but does require programmers to understand parallelism, locality, and bandwidth—general-purpose computing primitives.

More research is needed in domain-specific parallel algorithms, because most applications are sequential. Sequential algorithms are almost never appropriate for parallel systems. Expressing algorithms in such a way that they satisfy the key SPERC properties and are performance portable across different parallel hardware and generations of parallel hardware requires investment and research in new programming models and programming languages.

These programming models must enable expert, typical, and potentially naïve programmers to use parallel hardware effectively. Since parallel programming is extremely complex, the expertise necessary to effectively work in this realm is currently only within reach of the most expert programmers, and the majority of existing systems are not performance portable. A key requirement will be to create modular programming models that make it possible to encapsulate parallel software in libraries in such a way that (1) they can be reused by many applications and (2) the system adapts and controls the total amount of parallelism that effectively utilizes the hardware, without over- or undersubscription. Mapping applications, which use these new models, to parallel hardware will require new compiler, runtime, and operating system services that model, observe, and reason, and then adapt to and change dynamic program behaviors to satisfy performance and energy constraints.

Because power and energy are now the first-order constraint in hardware design, there is an opportunity for algorithmic design and system software to play a much larger role in power and energy management. This area is a critical research topic with broad applicability.

## 1.3 The Rise of Mobile Computing, Services, and Software

Historically, the x86 instruction set architecture has come to dominate the commercial computing space, including laptops, desktops, and servers. Developed originally by Intel and licensed by AMD, the commercial success of this architecture has either eliminated or forced into smaller markets other architectures developed by MIPS, HP, DEC, and IBM, among others. More than 300 million PCs are sold each year, most of them powered by x86 processors.[13] Further, since the improvement in capabilities of single-core processors started slowing dramatically, nearly all laptops, desktops, and servers are now shipping with multicore processors.

Over the past decade, the availability of capable, affordable, and very low-power processor chips has spurred a fast rise in mobile computing devices in the form of smartphones and tablets. The annual sales volume of smartphones and tablets already exceeds that of PCs and servers.[14] The dominant architecture is U.K.-based ARM, rather than x86. ARM does not manufacture chips; instead it licenses the architecture to third parties for incorporation into custom SoC designs by other vendors. The openness of the ARM architecture has facilitated its adoption by many hardware manufacturers. In addition, these mobile devices now commonly incorporate two cores, and at least one SoC vendor has been shipping four-core designs in volume since early 2012. Furthermore, new heterogeneous big- and small-core designs that couple a higher performance, higher power core with a lower performance, lower power core have recently been announced.[15] Multicore chips are now ubiquitous across the entire range of computing devices.

---

[12] In 2004, Google introduced the software framework, MapReduce, to support distributed computing on large datasets on clusters of computers.

[13] See http://www.gartner.com/it/page.jsp?id=1893523. Last accessed on February 7, 2012.

[14] See http://www.canalys.com/newsroom/smart-phones-over take-client-pcs-2011. Last accessed on February 7, 2012.

[15] See www.tegra3.org; http://www.reuters.com/article/2011/10/19/arm-idUSL5E7LJ42H20111019. Last accessed on June 25, 2012.

The rise of the ARM architecture in mobile computing has the potential to adjust the balance of power in the computing world as mobile devices become more popular and supplant PCs for many users. Although the ARM architecture comes from the United Kingdom, Qualcomm, Texas Instruments, and NVIDIA are all U.S.-based companies with strong positions in this space. However, the shift does open the door to more foreign competition, such as Korea's Samsung, and new entries, because ARM licenses are relatively inexpensive, allowing many vendors to design ARM-based chips and have them fabricated in Asia.

However, just as technical challenges are changing the hardware and software development cycle and the software-hardware interface, the rise of mobile computing and its associated software ecosystems are changing the nature of software deployment and innovation in applications. In contrast to developing applications for general-purpose PCs—where any application developer, for example, a U.S. defense contractor or independent software vendor, can create software that executes on any PC of their choosing—in many cases, developing software for mobile devices imposes additional requirements on developers, with "apps" having to be approved by the hardware vendors before deployment. There are advantages and disadvantage to each approach, but changes in the amount and locus of control over software deployments will have implications for what kind of software is developed and how innovation proceeds.

A final inflection point is the rise of large-scale services, as exemplified by search engines, social networks and cloud-hosting services. At the largest scale, the systems supporting each of these are larger than the entire Internet was just a few years ago. Associated innovations have included a renewed focus on analysis of unstructured and ill-structured data (so-called big data), packaging and energy efficiency for massive data centers, and the architecture of service delivery and content distribution systems. All of these are the enabling technologies for delivery of services to mobile devices. The mobile device is already becoming the primary personal computing system for many people, backed up by data storage, augmented computational horsepower, and services provided by the cloud. Leadership in the technologies associated with distributed cloud services, data center hardware and software, and mobile devices will provide a competitive advantage in the global computing marketplace.

Software innovations in mobile systems where power constraints are severe (battery life directly affects user experience) are predicted to use a different model than PCs, in which more and more processing is performed in the "cloud" rather than on the mobile device. A flexible software infrastructure and algorithms that optimize for network availability, power on the device, and precision are heralding a challenging ecosystem.

Fundamental to these technologies are algorithms for ensuring properties such as reliability, availability, and security in a distributed computing system, as well as algorithms for deep data mining and inference. These algorithms are very different in nature from parallel algorithms suitable for traditional supercomputing applications. While U.S. researchers have made investments in these areas already, the importance and commercial growth potential demand research and development into algorithmic areas including encryption, machine learning, data mining, and asynchronous algorithms for distributed systems protocols.

## 1.4 Summary and Implications

Semiconductor scaling has encountered fundamental physical limits, and improvements in performance and power are slowing. This slowdown has, among other things, driven a shift from the single microprocessor computer architectures to homogenous and now heterogeneous multicore processors, which break the virtuous cycle that most software innovation has expected and relied on. While innovations in transistor materials, lithography, and chip architecture provide promising opportunities for improvements in performance and power, there is no consensus by the semiconductor and computer industry on the most promising path forward.

It is likely that these limitations will require a shift in the locus of innovation away from dependence on single-thread performance, at least in the way performance has been achieved (i.e., increasing transistor count per chip at reduced power). Performance at the processor level will continue to be important, as that performance can be translated into desired functionalities (such as increased security, reliability, more capable software, and so on.) But new ways of thinking about overall system goals and how to achieve them may be needed.

What, then, are the most promising opportunities for innovation breakthroughs by the semiconductor and computing industry? The ongoing globalization of science and technology and increased—and cheaper—access to new materials, technologies, infrastructure, and markets have the potential to shift the U.S. competitive advantage in the global computing ecosystem, as well as to refocus opportunities for innovation in the computing space. In addition, the computing and semiconductor industry has become a global enterprise, fueled by increasingly competitive overseas semiconductor markets and firms that have made large and focused investments in the computing space over the last decade. The possibility of new technological approaches emerging both in the United States and overseas reinforces the critical need for the United States to assess the geographic and technological landscape of research and development focused on this and other areas of computer and semiconductor innovation.

# 2

# The Global Research Landscape

An assessment of the global landscape for research and development (R&D) of advanced computing—especially efforts to address the computing performance challenges outlined in Chapter 1—must include an examination of research efforts related to the technologies described in Chapter 1, including semiconductor devices and circuits, architecture, programming systems, and applications. Further, assessing the global competitiveness in these technologies requires examination of both advanced research and development as well as how successful commercialization of these technologies has been and for whom.

Research capability is a leading indicator for a nation's future technical competitiveness in science-intensive technological fields. For the purposes of this report, a nation's research capabilities include the education provided by and output from universities as well as the training by and output from industry and government laboratories. This chapter examines two broad indicators to assess national technological research capabilities and competitiveness: (1) commercialization of semiconductors, as well as computing hardware and software, technologies; and (2) bilateral trade revenues from U.S. exports and imports of advanced electronics and technology products critical to the computing performance challenges described in Chapter 1.

The committee also conducted a pilot study of a third indicator: national contributions of papers at top technical conferences. In computer science, papers presented at conferences are an important (and often underused) measure of research quality, in addition to journal articles. The committee analyzed authorship—specifically, authors' geographical locations—of papers at many of the top technical conferences in the four research areas most closely related to the challenges outlined in Chapter 1: semiconductor devices and circuits, computer architecture, programming systems, and applications. The pilot study included an analysis of data from papers presented at 2011 conferences in these four research areas and a similar analysis of conference papers from 1996–2011 to show recent changes in representation at these conferences. The complete results of the pilot study, along with a methodological overview and discussion of its limitations, are discussed in Appendixes F and G.

Section 2.1 provides a brief snapshot of some preliminary observations and insights that can be gleaned from the pilot study. Section 2.2 uses the conference publication data to examine how the international collaborative nature of these conference papers has changed over time. Section 2.3 provides a description of the global landscape in commercialization of semiconductor, as well as computing hardware and software, technologies. Section 2.4 presents an analysis of bilateral (U.S.-China, -Korea, -Taiwan, and -Japan) trade data for U.S. exports and imports of electronics and products specifically relevant to the computing challenges outlined in Chapter 1. Lastly, Section 2.5 examines China's growing role as a major consumer and supplier of semiconductors, as well as its contribution to the global semiconductor value chain.

## 2.1 Preliminary Observations from Pilot Study of Papers at Top Technical Conferences

The committee encountered some methodological challenges in its analysis of publication data from conference papers (for example, determining whether the location of a conference can introduce travel biases for researchers), making it difficult to draw concrete conclusions about the technological research capabilities of individual countries. There are some interesting observations about overall trends and emerging strengths, however, which can be made from this preliminary analysis.

*2.1.1. Current (2011) national and regional advanced computing research contributions at top technical conferences*

The committee's preliminary assessment of conference papers at selected technical conferences in 2011 indicates that the United States is strongly represented in each of the four research areas identified by the committee as critical for meeting the computing performance challenges outlined in Chapter 1 (semiconductor devices and circuits, architecture, programming systems, and applications), contributing more than half of all papers across each research area. Of these areas, the United States has the strongest representation in architecture research with no other individual nation contributing as significantly. These data are consistent with the historical U.S. strengths in commercial microprocessors, including Intel, AMD, and IBM, as well as former commercial microprocessors from DEC, HP, and others (see Table F-5). The committee notes, however, that the UK-based ARM processor ecosystem now dominates by processor shipment volume, largely based on smartphones and embedded devices.

Limited or no representation at architecture research conferences may suggest that some nations' universities and industry research institutions are not focused on mainstream computer architecture. For example, while Japan has activity and expertise in architecture research, notably the custom processors from Fujitsu that are in the K supercomputer, the data suggest its national research focus may lie in other areas such as advanced semiconductor and nanoscale devices and circuits (see Table F-4). As another example, Germany and the U.K., while poorly represented at architecture research conferences, have notable representation in advanced programming research (see Table F-6).

Several interesting observations can also be made about regional representation at these conferences. For example, while the United States maintains a significant lead over Europe and Asia in paper contributions at semiconductor and nanoscale devices conferences, its contributions in semiconductor circuits research are comparable to Europe and Asia. In programming systems and applications, the United States maintains a lead followed by Europe and distantly by Asia. See Figures F-2 through F-6.

*2.1.2 Time series assessment of national and regional advanced computing research contributions at top technical conferences*

Longitudinal analysis of conference data from 1996–2011 also provides insight into trends in national (see Tables F-8 through F-11) and regional (see Figures F-7 through F-11) contributions to advanced research. During this time, for the two conference series (IEDM and NANO) in the semiconductor devices area, the U.S. lead has remained relatively stable with the largest gains made by Taiwan and Belgium (IMEC). For the ISSCC conference series in semiconductor circuits research, the United States shows a moderate decline, in tandem with an overall broadening in international representation. In this area, the largest leaps were made by Korea, Taiwan, and the Netherlands. For the four conference series (ASPLOS, HPCA, ISCA, and MICRO) in architecture research, the United States has maintained a significant lead, with no major advances by any other nation or region.

For the five conference series (ECOOP, OOPSLA, PLDP, POPL, and PPoPP) in programming systems research, the U.S. lead has been challenged somewhat by increases in Europe by small but steady gains by Israel, Switzerland, and the UK (as well as by China, India, and Korea to a lesser degree). For the seven conference series (Eurographics, OSDI, SC, SIGGRAPH, SOSP, VLDB, and WWW) in applications research, U.S. representation has retained a stable lead over the 15-year period with no significant representation by other nations. While only representing a small percentage of papers in the applications research areas, China moved from no representation in 1996 to ~4 percent of conference papers in 2011.

Strong R&D investments by U.S. universities and industry laboratories over the last 15 years have yielded numerous innovations and have helped to sustain the United State's position as a lead contributor of conference papers across the four specific technology areas identified by the committee. Despite this fact, the U.S. position is now being challenged by increasing technical and manufacturing capabilities in Europe and, in particular, Asia. For example, while showing relatively few contributions to conference papers, China continues to make significant contributions to U.S.–China trade revenues (discussed in Section 2.4) and demonstrates increasing competitiveness in the global semiconductor value chain (discussed in Section 2.5). The committee expects that these trends will likely continue as nations make greater investments in domestic university and industry research, as well as through multinational, and increasingly global, commercial partnerships and international research collaborations.

## 2.2 Increased International Collaboration

Data on coauthored papers presented at several of the sampled conferences[1] discussed above and in Appendix F were used to examine how international collaborations have changed over time. In the network connectivity graphs[2] of Figure 2-1, the nodes (circles) represent individual countries, and the size of each node represents the number of papers produced by that country. The edges (lines connecting two circles) represent collaborations on coauthored papers, and the weight of each edge indicates the number of papers that share coauthorship between nations.

In each area except architecture, the network graphs show an increasing geographical diversity in research and a tremendous increase in international collaborations. The network graphs show that between 1996 and 2011, international participation and collaboration between the United States and other nations has dramatically increased. In the devices and circuits areas, many of the international collaborations come from work that spans multiple international sites within the same company. This trend toward greater collaboration across national boundaries will likely continue due to the increasing global investments in research by both nations and global industries.

International research collaborations in computer architecture have not increased dramatically, although more papers are being published as collaborations between U.S. and foreign researchers. The emergence of the ARM architecture in the mobile computing space provides impetus for foreign investment in architecture research, particularly in Europe, as European funding agencies prefer to invest in activities that are synergistic with European-based technologies.

Today, leading U.S. universities are linking to remote campuses in Asia and Europe and are describing themselves as "global universities." This trend, as well as the growing number of global companies, may have an impact on future U.S. competitiveness.

## 2.3 Commercialization of Technologies

This section provides a snapshot of the global landscape in the commercialization of semiconductor and computing hardware and software technologies using data from iSuppli, Gartner, the Hardware Top 100, and the Software Top 100.[3]

### 2.3.1 Semiconductor Commercialization

The committee began by analyzing revenues from the largest semiconductor, as well as computing hardware and software, companies. Table 2-1 shows the top 20 semiconductor companies, ranked by 2010 revenues[4] and includes companies that sell semiconductor components.[5] The chart shows the nation where the company is headquartered, its primary technology area, whether it has its own in-house fabrication capability, 2010 revenues in U.S. dollars, and the fraction of the global semiconductor market.

These top 20 companies account for a total of $197 billion, which is about two-thirds of the global semiconductor market. Of these top 20 companies, the United States accounts for 47 percent of revenue. Japan and Korea account for about 20 percent each, while Europe accounts for 10 percent. Historically, being a major semiconductor company required owning and operating significant semiconductor fabrication factories. However, the rising cost of deploying such facilities, both in R&D and capital investments, combined with the availability of "fab-for-hire" foundry services from companies such as Taiwan Semiconductor Manufacturing Corporation (TSMC), have given rise to an increasing number of fabless[6] semiconductor companies. Foundries such as TSMC have grown to be about 10 percent of the overall semiconductor component market (Gartner[7] estimate is U.S. $28.3

---

[1] Conferences included in each of the four technology areas are as follows: (1) Architecture: ASPLOS, HPCA, ISCA, and MICRO; (2) Programming: ECOOP, OOPSLA, PLDI, POPL, and PPoPP; (3) Applications: SIGGRAPH, SC, VLDB, and WWW; and (4) Semiconductor Devices and Circuits: IEDM and ISSCC.

[2] Coauthor networks were generated with the Science of Science (Sci$^2$) Tool: Sci2 Team (2009). Science of Science (Sci2) Tool. Indiana University and SciTech Strategies, http://sci2.cns.iu.edu.

[3] See www.isuppli.com; www.gartner.com; www.hardwaretop100.org; and www.softwaretop100.org.

[4] See http://www.isuppli.com/Semiconductor-Value-Chain/News/Pages/Intel-Reasserts-Semiconductor-Market-Leadership-in-2011.aspx. Last accessed on August 16, 2012.

[5] Companies that supply only fabrication services (such as TSMC with 2010 revenues of over $13 billion) are not included. Systems companies that design their own chips (such as Apple) are included in Table 2-1 below.

[6] Fabless semiconductor companies specialize in the design and sale of hardware devices and semiconductor chips, as opposed to device fabrication.

[7] "Semiconductor foundry revenue increased 40.5%, reaching $28.3 billion in 2010. Foundry fab utilization reached its peak in 3Q10 after several quarters of good growth. Leading-edge technologies (65 nm to 45 nm) have been in high demand from foundries, increasing in revenue contribution." Available at http://www.gartner.com/id=1634315. Last accessed on February 7, 2012.

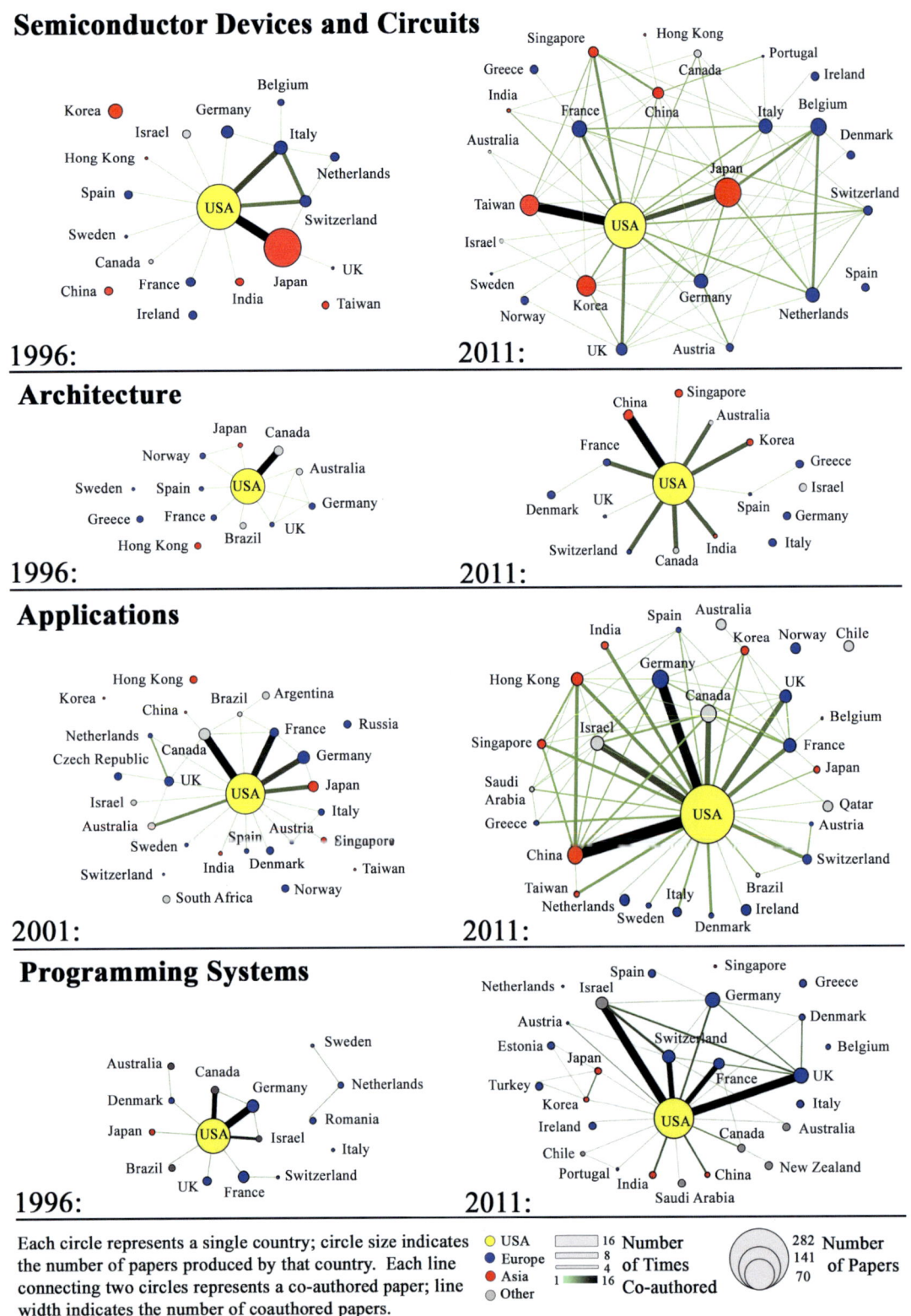

FIGURE 2-1 International conference collaboration networks. Data compiled from the following conferences: ASPLOS, HPCA, ISCA and MICRO (architecture); ECOOP, OOPSLA, PLDI, POPL, and PPoPP (programming systems); SC, SIGGRAPH, VLDB, and WWW (applications); and IEDM and ISSCC (semiconductor devices and circuits). Collaboration maps were generated using the Science of Science (Sci2) Tool available at http://sci2.cns.iu.edu.

TABLE 2-1 Top 20 Largest Semiconductor Companies (by revenue) in 2010

| Rank | Company | Country of Origin | Primary Market | Fab | $ Sales (U.S. millions) | % Market Share |
|---|---|---|---|---|---|---|
| 1 | Intel* | USA | microprocessors | yes | 40,394 | 13.2 |
| 2 | Samsung* | South Korea | memory, mobile SoCs | yes | 28,380 | 9.3 |
| 3 | Toshiba | Japan | memory | yes | 13,010 | 4.3 |
| 4 | Texas Instruments* | USA | DSP, mobile SOC | yes | 12,944 | 4.3 |
| 5 | Renesas | Japan | microcontrollers | yes | 11,893 | 3.9 |
| 6 | Hynix | South Korea | memory | yes | 10,380 | 3.5 |
| 7 | ST Microeletronics | France, Italy | memory, microcontrollers | yes | 10,346 | 3.4 |
| 8 | Micron | USA | memory | yes | 8,876 | 2.9 |
| 9 | Qualcomm* | USA | mobile SOC | no | 7,204 | 2.4 |
| 10 | Broadcom | USA | communication | no | 6,682 | 2.3 |
| 11 | Elpida | Japan | memory | yes | 6,446 | 2.1 |
| 12 | AMD* | USA | microprocessors, GPUs | no | 6,345 | 2.1 |
| 13 | Infineon | Germany | microcontrollers | yes | 6,319 | 2.0 |
| 14 | Sony | Japan | LCD, microprocessors | yes | 5,224 | 1.8 |
| 15 | Panasonic | Japan | microcontrollers | yes | 4,946 | 1.7 |
| 16 | Freescale | USA | microcontrollers | no | 4,357 | 1.4 |
| 17 | NXP | Netherlands | microcontrollers, mixed signal | yes | 4,028 | 1.3 |
| 18 | Marvell* | USA | mobile SOCs | no | 3,606 | 1.2 |
| 19 | MediaTek | Taiwan | communication | no | 3,553 | 1.2 |
| 20 | NVIDIA* | USA | GPUs, mobile SOCs | no | 3,196 | 1.0 |

Data compiled from isuppli's Preliminary Worldwide Ranking of the Top 20 Suppliers of Semiconductors in 2010 (www.isuppli.com).

billion foundry revenue out of roughly U.S. $300 billion overall semiconductor revenue). This trend has enabled startup companies to grow into large semiconductor companies, focused on design. While 13 of the top 20 on the 2010 list have their own semiconductor fabrication capability, 6 fabless semiconductor companies make the list, all of which are from the United States. The companies in the table that are marked by an asterisk design and sell multicore processors. Companies that sell multicore processors for PCs and servers (including graphics and high-performance accelerators) are Intel, AMD, and NVIDIA. Companies such as IBM and Oracle also design and sell multicore server processors, but have semiconductor revenues that place them outside the top 20. Companies that produce multicore processors for mobile devices such as cell phones and tablets include Samsung, Texas Instruments, Qualcomm, Marvell, and NVIDIA. As noted in Chapter 1, while the dominant instruction set in the PC and server space is x86 (Intel and AMD), the ARM instruction set dominates the mobile computing space. In addition to the mobile processor companies listed above, Apple designs its own multicore ARM-based processors for its mobile and tablet computers. The openness of the ARM architecture and ecosystem, along with ARM's focus on power efficiency, has led it to dominate in this fast-growing space.

It is equally important to assess how commercialization of these semiconductor technologies has changed over time. Table 2-2[8] ranks the largest semiconductor companies in 5-year intervals between 1995 and 2010. While mergers and acquisitions have changed the names of some of the companies, the country of origin still reflects the relative competitive stature of different nations and regions. In general, the United States has become more competitive in the semiconductor sectors. In 1995, 6 U.S. companies were in the top 20, representing 34 percent of the revenue of the top 20 companies. By 2010, 9 U.S. companies were in the top 20, representing 47 percent of revenue of the top 20 companies. While South Korea saw a drop in the number of companies in the top 20 from 3 to 2, the combined revenue share of Samsung and SK Hynix accounted for 19 percent of revenue of the top 20. Japan's representation in the top 20 also dropped, from 7 to 5, with its revenue share dropping even more precipitously, from 44 percent to 21 percent of the top 20.

*2.3.2 Computing Hardware and Software Commercialization*

In addition to assessing nations' competitive posture in the commercialization of semiconductor technologies, insight can also be gained from monitoring the world's largest computing hardware (including semiconductors, devices, and systems) and software companies. According to the Top 100 Research Foundation,[9] in 2010 the world's 20 largest hardware companies accounted for nearly U.S. $650 billion in annual revenue. Of just these top 20 companies, alone, the United States accounts for about 35 percent of total revenue, followed by Japan, South Korea, and Taiwan, each with about 19 percent. Europe's only entry is Nokia in Finland. China's Lenovo is a relatively recent entry to the global market, following their acquisition of IBM's laptop business in 2005.

A similar analysis finds that the world's 20 largest software companies account for more than U.S. $160 billion in revenue and nearly half of the overall U.S. $300 billion plus worldwide software market[10] in 2010 (see Appendix H). U.S.-based companies account for nearly 80 percent of revenues, with European companies accounting for about 15 percent. The only Asian country represented is Japan, with about 6 percent of the top 20 in revenue. The top U.S. companies, including Microsoft, IBM, Oracle, and HP, all have significant R&D investments in software and tools for parallel and multicore systems. The companies that produce game software all have core competence in parallel and multicore systems.

Information technology (IT) companies such as Google and Amazon do not appear in this list because their business models do not rely on selling software. However, they depend on a distributed parallel infrastructure that is now based on multicore systems. They are thus both producers and consumers of multicore hardware and software technology.

*2.3.3 Summary of Commercialization Landscape*

The degree to which indicators of national research capability and productivity, such as those discussed in Appendix F, are correlated with a nation's current commercial competitiveness, is a complex question, especially when the lag between research discovery and commercialization can be substantial and global information flow makes research results widely available. Similarly, the interplay between a country's research prowess and its educational systems affect global talent flow and retention in subtle and complex ways. These complexities underlie the longstanding questions about the interplay between basic research and technology commercialization, with broad implications for national, regional, and global economic policies.

Conversely, a nations' economic competitiveness may influence both its research capabilities and the ability of its companies to capitalize effectively on new research ideas. For example, if an industry can no longer translate the combination of government-funded basic research ideas and its own R&D investments into commercial successes with wide-enough profit margins, next-generation product development investments can become cost-prohibitive. This is akin to an economics argument that underlies Moore's Law—that the scaling rate parameter is significantly driven by the economics of internal investment and risk. For industries that can no longer make these investments or take the risks, residing at the leading edge of technology is no longer a viable business model and new strategies are required to remain competitive.

In considering a nation's ability to commercialize technological investments, it is important to recognize that most of the world's largest semiconductor, hardware, and software companies are global in nature, with R&D and manufacturing facilities worldwide, along with a complex set of technology cross-licensing agreements and supply chain interdependencies. Very

---

[8] Reported revenues for each company may not be independent; for example, due to the outsourcing of manufacturing across companies, as well as cross-licensing and use of intellectual property.

[9] See http://www.hardwaretop100.org/. Last accessed on June 16, 2012.

[10] See http://www.softwaretop100.org/. Last accessed on June 16, 2012.

# THE GLOBAL RESEARCH LANDSCAPE

TABLE 2-2 Largest Semiconductor Companies by Revenue (1995-2010)

| Company | Country of Origin | 1995 Rank | 2000 Rank | 2005 Rank | 2010 Rank |
|---|---|---|---|---|---|
| Intel | USA | 1 | 1 | 1 | 1 |
| Samsung | South Korea | 6 | 4 | 2 | 2 |
| Toshiba | Japan | 3 | 2 | 4 | 3 |
| Texas Instruments | USA | 7 | 3 | 3 | 4 |
| Renesas | Japan | N/A | N/A | 7 | 5 |
| SK Hynix | South Korea | N/A | 14 | 11 | 6 |
| STMicroeletronics | France, Italy | N/A | 6 | 5 | 7 |
| Micron | USA | 18 | 10 | 12 | 8 |
| Qualcomm | USA | N/A | N/A | 16 | 9 |
| Broadcom | USA | N/A | N/A | 20 | 10 |
| Elpida | Japan | N/A | N/A | N/A | 11 |
| AMD | USA | N/A | 16 | 15 | 12 |
| Infineon | Germany | N/A | 8 | 6 | 13 |
| Sony | Japan | N/A | 20 | 13 | 14 |
| Panasonic | Japan | N/A | N/A | N/A | 15 |
| Freescale | USA | N/A | N/A | 10 | 16 |
| NXP | Netherlands | 11 | 9 | 9 | 17 |
| Marvell | USA | N/A | N/A | N/A | 18 |
| MediaTek | Taiwan | N/A | N/A | N/A | 19 |
| NVIDIA | USA | N/A | N/A | N/A | 20 |
| NEC | Japan | 2 | 5 | 8 | N/A |
| Matsushita | Japan | 13 | 17 | 14 | N/A |
| Sharp | Japan | 19 | 19 | 17 | N/A |
| Rohm | Japan | N/A | N/A | 18 | N/A |
| IBM Microelectronics | USA | 12 | 18 | 19 | N/A |
| Motorola | USA | 5 | 7 | N/A | N/A |
| Mitsubishi | Japan | 9 | 11 | N/A | N/A |
| Hitachi | Japan | 4 | 12 | N/A | N/A |
| Agere | USA | N/A | 13 | N/A | N/A |
| Fujitsu | Japan | 8 | 15 | N/A | N/A |
| Hyundai | South Korea | 10 | N/A | N/A | N/A |
| SGS Thompson | France, Italy | 14 | N/A | N/A | N/A |
| Siemens | Germany | 15 | N/A | N/A | N/A |
| LG | South Korea | 16 | N/A | N/A | N/A |
| Sanyo | Japan | 17 | N/A | N/A | N/A |
| National Semiconductor | USA | 20 | N/A | N/A | N/A |

Data compiled from www.isuppli.com (2000, 2005, 2010) and www.gartner.com (1995).

few, if any products are designed, manufactured, and sold entirely within the borders of a single country.

For example, nearly all Taiwanese companies maintain manufacturing facilities in China, as does Intel and other U.S.-based companies. Many of these companies also operate assembly and test factories in lower-cost countries such as Vietnam, Malaysia, Costa Rica, and others. Furthermore, many companies on the list outsource manufacturing to other companies on the list. In particular, Foxconn, Quanta Computer, and Compal Electronics each manufacture systems on behalf of companies such as Toshiba, Dell, HP, and Apple. This interconnectedness of international economies underscores the need for researchers, as well as policy makers, to maintain a global awareness of not only emerging research capabilities, but also of successes in the commercialization of semiconductor, hardware, and software technologies.

In the last decade, Asia has gained an increasing role in the commercialization of technologies, particularly in manufacturing. In the areas centered on design (as exemplified by the U.S. fabless semiconductor industry), the United States still leads in both hardware and software. However, other nations seek to climb the value chain from manufacturing to integrated system design. Samsung's investment in its own system-on-a-chip designs are but one example of that type of activity.

In the following section, U.S. Census Bureau trade data on advanced technology products are examined to provide a view of how different nations transform innovation from advanced research investments into commercially successful products.

## 2.4 Growing Complexity in IT Trade – Tracing Shifts in International Competitiveness

Trade data provide additional information on the global R&D landscape in advanced computing technologies and products, and on potential future shifts in competitive advantages.

*2.4.1 U.S. Census ATP Trade Data at the 10-digit Level for Information and Communications (Code 4) and Electronics (Code 5)*

Analysis of U.S. Census Bureau Advanced Technology Product (ATP) trade data allows a closer look at changing patterns of trade in Code 4 information and communications technology (ICT) products and Code 5 electronics, including integrated circuits (ICs), products[11] between the United States and China, Taiwan,

Japan, and Korea. In addition, the ATP trade data may also provide useful proxy indicators on the development of technological capabilities in those four countries.

While U.S. ATP exports fared better than other U.S. exports during 2009,[12] the recession induced a great deal of volatility for information technology and electronics ATP exports. For example, U.S. electronics (including integrated circuits) exports fell by 27 percent in 2009 and then increased by 23 percent in 2010. The same volatility can be seen for U.S. ATP exports to Asia. After declining by 15 percent in 2009, U.S. ATP exports to Asia grew by a record 23 percent in 2010, driven by the rapid growth in both electronics and ICT exports. Here, China emerges as the most important growth determinant of U.S. ATP exports, electronics in particular. In fact, in 2009, electronics accounted for roughly half of U.S. ATP exports to China. It is important to emphasize that the focus increasingly is on semiconductors (~90 percent of U.S. electronics exports to China) intended for use in China's manufacturing plants.[13]

*2.4.2 Trade analysis of 'Advanced Computing' (AC) ATP exports/imports*

As an additional metric for assessing international competitiveness, U.S. Census Bureau ATP trade data at the 10-digit level was examined to quantify changes in ICT and electronics exports to (and imports from) China, Korea, Taiwan, and Japan between 2006 and 2010.[14] In

---

[11]The US Census Bureau defines "information and communications" (Code 4 of its ATP trade database) as products that are able to process increased volumes of information in shorter periods of time. This includes central processing units, all computers, and some peripheral units such as disk drive units and control units, along with modems, facsimile machines, and telephonic switching apparatus. Examples of other products included are radar apparatus and communication satellites. Code 5 (electronics) concentrates on recent design advances in electronic components (with the exception of optoelectronic components) that result in improved performance and capacity and, in many cases, reduced size. Products included are integrated circuits, multilayer printed circuit boards and surface-mounted components such as capacitors and resistors.

[12]D. Hill, September 2011, *U.S. Exports of Advanced Technology Products Declined Less than Other U.S. Exports in 2009*, InfoBrief, National Center for Science and Engineering Statistics, National Science Foundation. The U.S. Census Bureau defines ATP trade to consist of advanced materials, aerospace, biotechnology, electronics, flexible manufacturing, information and communication technology (ICT), life science, optoelectronics, nuclear technology, and weapons. Four of these 10 categories (i.e., aerospace, electronics, ICT, and life science) together accounted for 85 percent of U.S. ATP exports in 2010.

[13]A. Hammer, R. Koopman, A. Martinez, 2009, *U.S. Exports of Advanced Technology Products to China*, U.S. International Trade Commission, October, No. RN-2009-10E.

[14]This time period allows for a consideration of the effects of the 2008-2009 global recession.

particular, the committee selected from the Code 4 and 5 export-import data a narrower set of product groups directly relevant to the computer and semiconductor R&D and commercial ecosystem.[15] These focused product groups are referred to as Code 4 AC (advanced computing) and Code 5 AC. These AC products are technologically more complex than the rest of the Code 4 and Code 5 product groups, and hence may pose higher entry barriers for latecomers like China (discussed in Chapter 3).

Figure 2-2 compares the growth of U.S. Code 4 AC and 5 AC exports with that of all U.S. Code 4 and 5 exports. Between 2006 and 2010, U.S. Code 4 AC (blue triangles) exports grew four times faster than all Code 4 (blue circles) exports. In contrast, all U.S. electronics exports were negatively affected by the global recession, with Code 5 AC (red triangles) and Code 5 (red circles) exports falling ~11 percent and ~14 percent, respectively.

For U.S. imports, Figure 2-3 shows that U.S. Code 4 AC imports (blue triangles) grew more than twice as fast as all Code 4 imports (~68 percent compared to ~28 percent) between 2006 and 2010. In contrast, both U.S. Code 5 and 5 AC imports show relatively flat growth (± ~1 percent). During this time, the shares of both U.S. Code 4 AC and 5 AC exports in all Code 4 and 5 exports showed a slight increase of ~3-4 percent. The same is true for U.S. Code 4 AC and 5 AC imports, with the share of Code 4 AC imports in all Code 4 imports showing a larger increase of ~7 percent.

### 2.4.3 Changing Patterns of Trade

Figure 2-4 shows changes in total values of U.S. Code 4 and 4 AC exports to four leading Asian countries—China, Japan, Korea, and Taiwan—between 2006 and 2010. In 2007, China became the largest overseas market for U.S. Code 4 and 4 AC exports, followed by Japan, Korea, and Taiwan. Since then, the export markets of Korea, Taiwan, and Japan, in particular, have all declined (the most dramatic being a ~39 percent decline by Japan). In contrast, China has experienced remarkable growth as a market for U.S. exports of Code 4 and, especially, 4 AC products (~86 percent growth over 4 years). As a result, China held the largest share (nearly half) of all U.S. Code 4 AC exports to Asia in 2010, with Japan in second place (~34 percent). However, Japan's relatively stagnant export market is unlikely to be a challenger (at least in the near term) as China emerges as a leading market for more sophisticated U.S. ICT exports.

Figure 2-5 shows that China has also become an important market for U.S. Code 5 AC exports, well ahead of Taiwan, Korea, and Japan. Even during the global recession, China's market for U.S. electronics exports, including semiconductors, continued to increase, while each of the other Asian markets experienced significant declines, in particular Korea and Japan with more than 40 percent reductions. By 2010, China held the largest share (nearly half) of all U.S. electronics exports (both Code 5 and 5 AC) to Asia, followed by Taiwan (~25 percent) and Korea (~20 percent). As such, China has established itself as a leading market for sophisticated U.S. electronics exports, and especially for semiconductors.

Figure 2-6 shows that China clearly dominates as a major source of Code 4 AC products, well ahead of Korea (which overtook Japan in 2008), Japan, and Taiwan (by more than an order of magnitude). Between 2006 and 2010, U.S. Code 4 AC imports from China grew by almost 130 percent. More importantly, by 2010 China held a ~95 percent share of all Code 4 AC exports from Asia to the United States, establishing its role as a major source of U.S. imports of sophisticated ICT products. While they do not pose any perceivable threat to China's lead, it is worth noting that an increasing share of U.S. Code 4 imports from Korea and Taiwan,[16]—but not Japan, which has experienced negative growth—now constitute more sophisticated Code 4 AC products.

Figure 2-7 shows that, despite the trade disruption resulting from the global recession, Taiwan remains the leading source of Code 5 AC imports to the United States, ahead of Korea, Japan, and China. Historically, China has lagged behind the rest of Asia as a source of U.S. Code 5 and 5 AC imports. However, it is worth emphasizing that, since 2009, Code 5 AC imports to the United States from China have grown much faster (~72

---

[15]From the Code 4 and Code 5 data, the committee analyzed import and export products associated with advanced computing. In particular, the committee focused on categories that included (1) products associated with the implementation of integrated circuits and (2) memory and logic-integrated circuits themselves. The committee did not include discrete electronic components (e.g., diodes and amplifiers), display technologies, low-frequency integrated circuits, printer technologies, magnetic storage, and radio and telecommunication technologies.

[16]In 2010, U.S. imports of Code 4 AC products from Korea and Taiwan grew by ~1,328 percent and ~98 percent, respectfully. However, high growth rates should not necessarily be associated with high export-import values. For example, while Korea exhibits a high growth rate of Code 4 AC products to the United States between 2006 and 2010, the total value of these exports remains very low (increasing from U.S. $30 million to $470 million), compared with China (U.S. $16.9 billion to $38.8 billion).

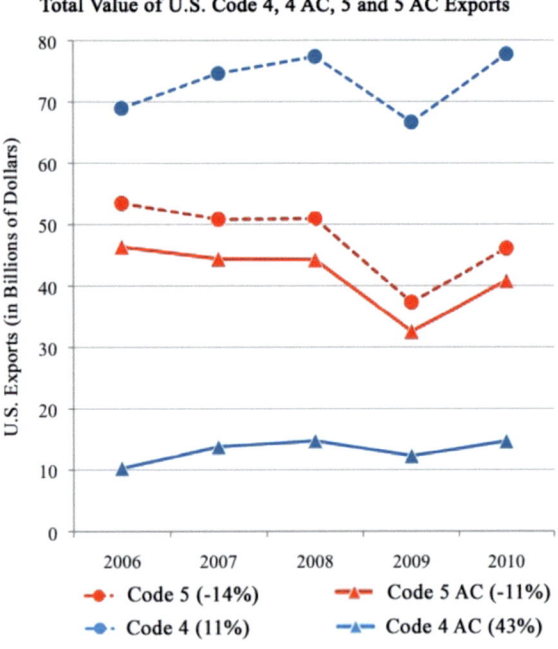

FIGURE 2-2 Total value of U.S. Code 4, 4 AC, 5, and 5 AC exports. Data compiled from U.S. Census Bureau Advanced Technology Product trade data.

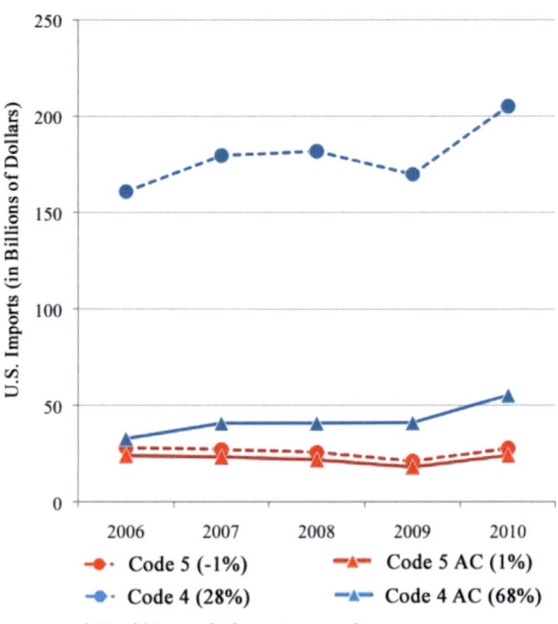

FIGURE 2-3 Total value of U.S. Code 4, 4 AC, 5, and 5 AC imports. Data compiled from U.S. Census Bureau Advanced Technology Product trade data

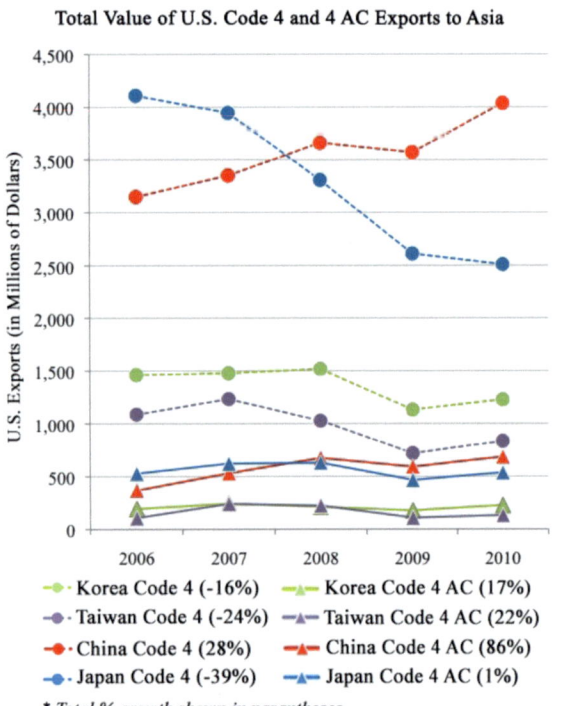

FIGURE 2-4 Total value of U.S. Code 4 and 4 AC exports to Asia. Data compiled from U.S. Census Bureau Advanced Technology Product trade data.

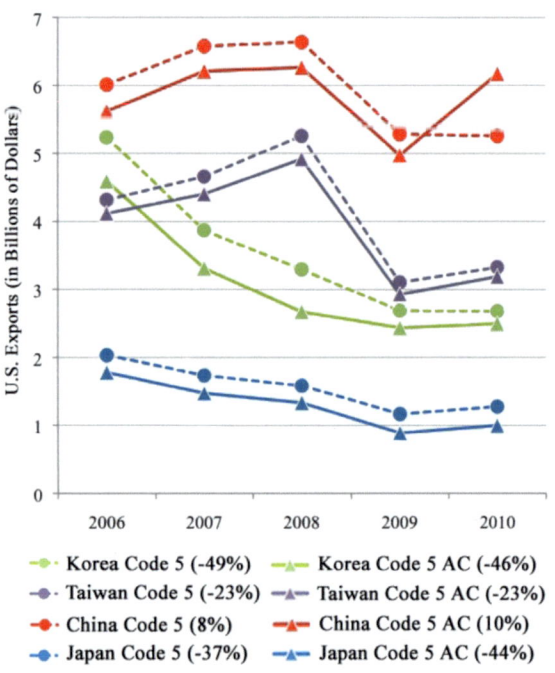

FIGURE 2-5 Total value of U.S. Code 5 and 5 AC exports to Asia. Data compiled from U.S. Census Bureau Advanced Technology Product trade data.

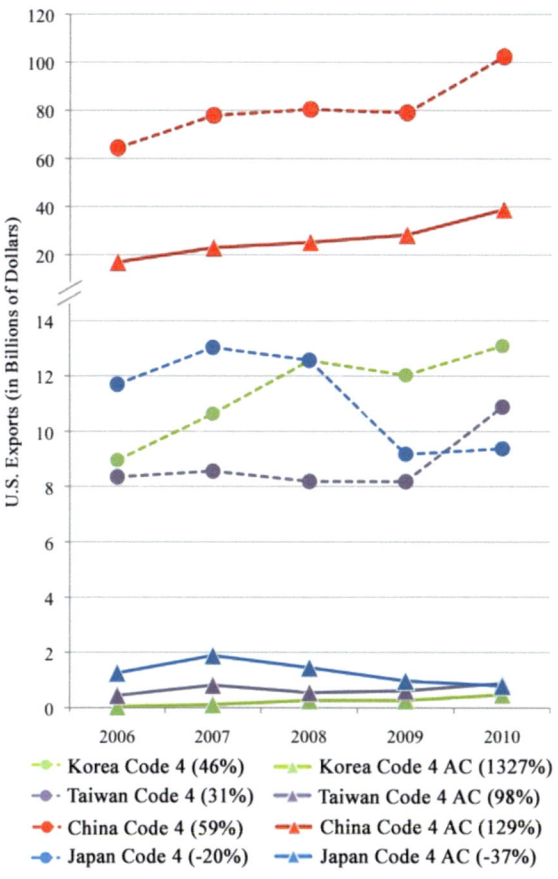

FIGURE 2-6 Total value of U.S. Code 4 and 4 AC imports from Asia. Data compiled from U.S. Census Bureau Advanced Technology Product trade data.

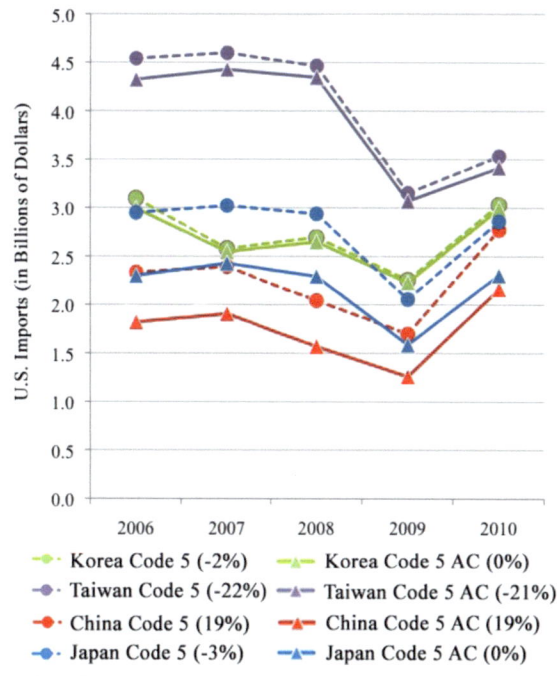

FIGURE 2-7 Total value of U.S. Code 5 and 5 AC imports from Asia. Data compiled from U.S. Census Bureau Advanced Technology Product trade data.

percent) than any of the other three Asian countries. As a result, China now has a 20 percent share of all U.S. Code 5 AC imports from Asia, suggesting that China is making continuous progress as an exporter despite its persistent weakness as a semiconductor producer.

*2.4.4 Summary of Trade Data Analysis*

The overall picture that emerges from the analysis of trade data is that the United States has kept its leading position as a supplier of leading-edge semiconductors, as reflected by Code 5 and 5 AC export data. China is also increasingly becoming a major market for both Code 5 and 5 AC products. Most U.S. exports of semiconductors to China end up in Chinese ICT products. The trade data also suggest that China is exhibiting increased competitiveness as both a consumer and supplier of Code 4 and 4 AC products.

Another important finding is that shifts in competitiveness are very pronounced among the four leading Asian countries. While these shifts differ across product markets, China consistently leads the group, except in being a supplier of Code 5 and 5 AC products (though it is growing in that market, as well).

Finally, with regard to China's position in the global semiconductor value chain, the trade data analysis suggests that while the China market increases in importance, its domestic semiconductor industry continues to play a secondary role. Although China has growing strengths in the O-S-D (optoelectronics-sensor-discrete) industry and in the SPA&T (semiconductor packaging, assembly, and test) industry, these are somewhat secondary markets that do not define future technology trajectories.

In summary, the United States cannot afford to be complacent. There is no doubt that over time China's position in the global semiconductor value chain will keep improving. The United States needs to be prepared for a long-term shift in competitiveness.

## 2.5 China's Position in the Global Semiconductor Value Chain

So far, three broad metrics have been discussed—preliminary pilot study data on national representation at technical conferences (i.e., authorship of conference papers); revenues generated by leading semiconductor, hardware, and software companies; and analyses of advanced electronics and ICT imports from and exports to Asia—to assess national technological leadership and competitiveness—each of which has indicated that the United States continues to maintain a leadership position.

However, these metrics also shed insight into the potential for competitor nations, such as China, to now meet or surpass U.S. technological capabilities. For example, while Table F-4 suggests that China's research contributions (via conference papers) are low relative to the United States and other leaders like Japan, Section 2.4 shows that China is exhibiting increased competitiveness as a strong consumer and supplier of advanced electronics and ICT products. China's rapidly growing semiconductor market continues to transform the semiconductor industry worldwide (both geographically and economically). Accordingly, many questions arise as to whether and how these emerging changes may affect the global semiconductor value chain.

The following sections examine China's growing role as a major consumer and supplier of semiconductors, as well as its contribution to the global semiconductor value chain. The following description relies heavily on the PricewaterhouseCoopers (PwC) report, *Continued growth – China's impact on the semiconductor industry – 2011 update*.[17]

### 2.5.1 China has become a Major Consumer of Semiconductors

Over the last decade, China's semiconductor market[18] has grown at an incredible 24.8 percent compounded annual growth rate (CAGR), far outpacing the worldwide semiconductor market (3.9 percent CAGR). In 2010 alone, China's semiconductor market grew by ~30 percent to U.S. $132 billion, accounting for more than 40 percent of the worldwide market. Much of this growth is driven by two factors: (1) a significant portion of consumed semiconductors are incorporated into final products that are assembled in China and then exported for sale elsewhere (such as in the United States, the European Union, Japan, and India) and (2) electronic products produced in China have higher semiconductor content (~27 percent) than the worldwide average (~19 percent).

To fairly assess the implications of China's large share of the worldwide semiconductor market, it is important to recognize that, unlike in the United States, China's semiconductor market is dominated by multinational and global semiconductor companies. For example, the 10 largest suppliers of semiconductors to China are not Chinese-owned companies and account for a combined 47 percent share of China's semiconductor market. However, although no Chinese-owned semiconductor companies are among the top 10 suppliers to either the worldwide or Chinese market, China's domestic semiconductor market has experienced significant growth, increasing from U.S. $10 billion in 2003 to U.S. $46 billion in 2010, representing more than 27 percent of worldwide semiconductor market growth.

Today, China's domestic consumption of semiconductors accounts for more than 15 percent of the worldwide market. This suggests that Chinese original equipment manufacturers (OEMs) may play a more prominent role in the future in shaping the parameters of semiconductor designs.

### 2.5.2 China's Semiconductor Manufacturing Industry

In 2010, China's semiconductor industry[19] experienced record growth (~30 percent) with revenues that account for ~8–11 percent of the worldwide semiconductor industry (compared with ~2 percent in 2000). Similar to its largest semiconductor suppliers, China's largest semiconductor manufacturing enterprises are multinational integrated device manufacturers (IDMs). Various aspects of the industry have been demonstrating significant growth.[20]

China's IC design industry has become one of the fastest growing sectors in China's semiconductor

---

[17] PwC, 2011, *Continued growth: China's impact on the semiconductor industry – 2011 update*. Available at http://www.pwc.com/gx/en/technology/assets/china-semiconductor-report-2011.pdf. Last accessed January 27, 2012.

[18] China's semiconductor market refers to the value of all semiconductor devices consumed in China by EMSs (electronics manufacturing service providers), OEMs (original equipment manufacturers), and ODMs (original design manufacturers).

[19] China's semiconductor industry refers to the sum of all reported revenues of all semiconductor manufacturers in China, including IC design, IC manufacturing and wafer foundries, IC packaging and test, and O-S-D companies.

[20] See China Semiconductor Industry Association (年中国半导体十大企业 2010). Last accessed on February 23, 2012.

industry, generating revenues of U.S. $5.4 billion (46 percent CAGR) in 2010, up from U.S. $178 million in 2001.[21]

In 2010, China's semiconductor packaging, assembly, and test (SPA&T) sector also experienced positive growth (~27 percent) with revenues of U.S. $9.3 billion, representing 20 percent of worldwide SPA&T facilities.[22] China's O-S-D sector, in particular its light-emitting diode industries, experienced similar growth with revenues more than twice that of its SPA&T sector (U.S. $23.4 billion). In addition, revenues generated in 2010 by China's wafer foundries grew by more than 45 percent, accounting for ~11 percent of worldwide foundry revenues.

*2.5.3 Contribution to the Global Semiconductor Value Chain*

To assess China's position in the global semiconductor value chain, it is important to assess China's strengths and weaknesses along each step of the value chain. Table 2-3 reports disaggregated semiconductor value chain revenue generated both by China and worldwide. The data illustrates that China currently acts primarily as a semiconductor consumer, accounting for ~37 percent of the worldwide semiconductor value chain. While the majority of these semiconductors consumed in China were ultimately exported for sale outside of China, more than one-third were used in electronic products consumed within China.

In contrast, China's contribution as a semiconductor producer, that is, sales, only accounts for ~8 percent (total sales in China divided by total worldwide revenue) of the worldwide value chain. Although its aggregated contributions as a semiconductor provider remains low, China is also a strong contributor to worldwide discrete device revenues and continues to develop its IC design capabilities.

*2.5.4: Summary of China's Position in the Global Semiconductor Value Chain*

In the last several years, China has steadily increased its position in the global semiconductor value chain—particularly as a consumer of semiconductor devices. China's IC design industry has also made significant gains. However, numerous challenges remain that have

---

[21] See China High-Tech Industry Development Almanac (中国高技术产业发展年鉴 2010). Last accessed on February 23, 2012

[22] See semi.org.cn (周斌). Last accessed on February 23, 2012.

Table 2-3 China's 2010 Contribution to Worldwide Semiconductor Value Chain Revenue (in Billions of U.S. Dollars)

| | Worldwide Revenue | China Sales | China Consumption |
|---|---|---|---|
| Electronic Design Automation | 4.2 | N/A | 0.31 |
| Semiconductor Intellectual Property | 1.5 | N/A | 0.12 |
| Equipment | 39.5 | 0.10 | 3.63 |
| Materials | 43.6 | 0.43 | 4.15 |
| IDMs | 224.7 | 18.9 | 99.6 |
| Fabless | 73.6 | 5.4 | 32.4 |
| Foundries | 30.2 | 3.2 | 13.4 |
| SATS | 23.6 | 9.1 | 10.4 |
| Total | 440.9 | 36.6 | 164.01 |

Adapted from the PwC report: *Continued growth: China's impact on the semiconductor industry –2011 update*, including source material from CSIA, EDAC, Gartner Dataquest, GSA, and SEMI (available at www.pwc.com).

the potential to reduce China's overall position in the value chain, as well as to alleviate concerns that the semiconductor value chain is threatened in the near term. While China has certainly emerged as what might be termed the dominant global factory for IT equipment, all products manufactured in China could also be manufactured elsewhere if there was an interruption in trade with China. In addition, the overwhelming majority of suppliers to China's semiconductor market are foreign companies.

While China's IC design industry continues to experience strong growth, it lags behind the United States, Japan, Taiwan, and Korea in terms of process technology and design line width. This is partly a result of (1) lacking or technologically inferior Chinese suppliers of electronic design automation tools and software and domestic licensors of IC design-related intellectual property and (2) significant supply-side constraints (e.g., intense competition and price wars) that have bankrupted many of China's domestic IC houses. In addition, a narrow focus on low- and middle-end consumer products threatens to constrain the growth of China's IC design industry. By fueling its dependence on mature and relatively standardized products, China limits its R&D and semiconductor capability developments. Similarly, it is questionable whether China can sustain its increasing share of total worldwide wafer production as its leading foundries (e.g., SMIC, Shougang NEC Electronics, and HeJian Technology) have experienced dramatic revenue declines in 2010. While it continues to lead in the number of new wafer fabrications, these plants use older technology.

## 2.6 Concluding Remarks

This chapter provides two measures for assessing advanced research efforts related to the key advanced computing technologies described in Chapter 1, as well as for assessing global competitiveness in these technologies. These two measures include (1) descriptions of the global landscape of the commercialization of semiconductor, computing hardware, and software technologies; and (2) bilateral trade data analysis of electronics and ICT ATP products with a focus on technologies specifically relevant to the computing challenges outlined in Chapter 1.

Preliminary observations from a pilot study of a third possible measure, conference publication data, indicate that the United States has maintained its position as a strong contributor of research papers at the technical conferences sampled by the committee over the last 15 years, with particular strengths in the area of architecture research. Early results from the pilot study also demonstrate the value of a more focused examination of a nation's technology-specific paper contributions (say, compared to a bulk analysis across all advanced computing sectors). The longitudinal data analyses also provide a starting point for identifying trends in national and regional participation in specific technology areas. While these early data suggest a strong U.S. position, a more thorough investigation is necessary.[23]

The advanced technology product trade data analyses indicate that China is increasingly becoming a major market for advanced electronics products. China also is exhibiting increased competitiveness as both a user and supplier of ICT products; its domestic semiconductor industry also continues to grow. In addition, Taiwanese manufacturing and assembly are increasingly being transferred to China. While the United States has maintained—and is likely to continue in the near term—its leading position as a supplier of leading-edge semiconductor designs, it has a more minor position in semiconductor manufacturing.

Though conference data from the pilot study suggest that China currently lags behind the United States in leading research capabilities (though it is growing, especially in the applications research areas), it is important to consider other indicators of China's research capabilities—for example, the technology transfer through U.S. education of Chinese foreign nationals. By sending its best students to top U.S. research universities, China can capitalize on the "value added" by American education and bootstrap its manufacturing and design prowess without these capabilities showing up in publication data.

China is already a major consumer of ICT products, which is increasingly shaping product expectations and standards; over time, China's position in the global semiconductor value chain will continue to improve. Thus, the United States cannot be complacent and needs to be prepared for this long-term shift in competitiveness.

---

[23]For example, as discussed in Appendix F, future iterations of this analysis would ideally take into consideration all papers relevant to the computing performance challenges outlined in Chapter 1, whether published at conferences or in traditional journals, weighted by citations and impact factors, as well as expert judgment.

# 3

# Innovation Policy Landscape – Comparative Analysis

Fundamental physical limitations in semiconductor scaling have slowed future expectations of improved, single-processor computing performance on which all sectors of society in the United States—and around the world—have relied. Many of these technological challenges at the frontiers of device design, computer architecture, and parallel programming methodologies were described in Chapter 1. While many short-term technological fixes have led to recent computing performance improvements, no silver bullets have emerged to reclaim the steady exponential computing performance gains once achieved by successive generations of single microprocessor computer systems.

How, then, will the next generation of semiconductor, computer architecture and programming breakthroughs come about? What types of policies and institutions, whether public, private or partnerships, will be required to bring about the technological innovation necessary for next-generation hardware devices, system architectures, and programming systems? To address these questions, it is important to understand the role innovation policies have played in supporting U.S. innovation and, in the context of this report, the policies that strengthen, sustain, and/or erode the innovative capabilities of the computer and semiconductor industries.

Innovation policies differ across countries, industries, and technologies. Countries differ in their levels of development and in their economic institutions, and hence pursue quite different approaches to innovation policy. The United States model, for instance, has largely been based on a belief that market forces (which include government and defense as consumers that demand leading-edge technology) and the private sector should play a primary role in innovation, backed by government investment in basic academic research. In contrast, the European Union and emerging economies such as China, Korea, and Taiwan rely much more on the government to articulate strategic objectives and key parameters.

In the United States, there is a widespread expectation that government-centered innovation systems will "naturally" converge to a U.S.-style market-led system. However, comparative research on national innovation policies suggests that this convergence is limited.[1] In addition, innovation policies change over time, even within the same country. As Charles Vest emphasizes, the American innovation system has a long tradition of highly decentralized, market-driven innovation networks, where the government historically played a role primarily at the local level.[2] However, as ubiquitous globalization disaggregates manufacturing, product development, and research, it is not yet clear which policies will best support future innovation in the United States.[3]

To understand how requirements for innovation policy differ across industries and technologies, it is

---

[1] See R. R. Nelson, 1993, ed., *National Innovation Systems. A Comparative Analysis,* Oxford University Press, New York. For an analysis of the persistent diversity of China's and America's innovation and standards policies, see D. Ernst, 2011, *Indigenous Innovation and Globalization: The Challenge for China's Standardization Strategy*, University of California Institute on Global Conflict and Cooperation, La Jolla, CA, and East-West Center, Honolulu, HI, 123 pp.

[2] C. Vest, 2011, "Universities and the U.S. Innovation System," in C. W. Wessner, ed., *Building the 21st Century. U.S.-China Cooperation on Science, Technology, and Innovation*, Washington, D.C.: The National Academies Press, pp. 70-73.

[3] D. C. Mowery, 2009, Plus ca change: Industrial R&D in the "third industrial revolution," *Industrial and Corporate Change*, (18):1, pp. 1-50.

useful to consider how innovations differ—in the complexity of the infrastructure and capabilities required to foster and implement them. Furthermore, the demands of innovation policy differ, depending on the nature and the intensity of innovation barriers that constrain the deployment of new ideas, inventions, and discoveries into commercially successful products, services, and business models.

Today, the effects of globalization extend across all stages of the value chain, including engineering, product development, and applied and basic research. This has resulted in an increase in the organizational and geographic mobility of knowledge.[4] However, the new geography of knowledge is not a flatter world where technical change and liberalization spread the benefits of globalization rapidly and equally. Instead, even mature and established technology and manufacturing leaders now face competition from a handful of new—yet very diverse and intensely competitive—manufacturing and research and development hubs around the world.[5] Therefore, the United States can learn a great deal by looking at the strengths and weaknesses of alternative information technology (IT) innovation policies in other nations.[6] An analysis of these diverse approaches to innovation policy is shaped by issues such as: the range of policy options that have been pursued, how policy approaches differ, how these differences affect innovation capacities, and how innovation policies pursued elsewhere affect the global supply chain.

This chapter examines the strengths and weaknesses of different innovation strategies, policy tools, and institutional arrangements implemented in countries that are potentially important players in the development of computing devices, technologies, and products. While U.S. innovation strategies have primarily relied on market forces and the private sector, it is important to understand the varied and complex factors that drive the evolution of different national innovation ecosystems. For example, countries such as China and Taiwan have relied on top-down government leadership to define strategic objectives and key parameters of innovation programs. Another variant of innovation policy can be found in the European Union's recent push toward new forms of cross-border coordination of innovation markets and infrastructures.

Section 3.1 provides a history of the U.S. semiconductor industry and examines how America's decentralized market-driven innovation system has led to where the United States is today. Section 3.2 looks at China's indigenous innovation policy, especially its recent Strategic Emerging Industries (SEI) Program. Section 3.3 examines the evolving role of Taiwanese innovation policies to support low-cost and fast innovation through domestic and global innovation networks. Section 3.4 looks at Korea's coevolution of international and domestic knowledge linkages. Section 3.5 examines the European Union's recent efforts to develop an integrated innovation strategy and its recent Key Enabling Technologies (KETs) Program. Section 3.6 provides concluding remarks and policy implications.

## 3.1 Development of the U.S. Computer and Semiconductor Industry

### 3.1.1 Historical Context

Several factors influence the range and type of policy options available to nations to promote and manage development and competitiveness in the semiconductor, computer architecture, and software programming arenas. Among those factors historically dominating U.S. policy considerations are

- The economic importance of semiconductors and computing in the U.S. national economy;
- The economic importance of closely related U.S. industries (e.g., telecommunications, consumer electronics, military and aerospace);
- The outlook on the U.S. federal budget, the climate for public and private investment, the employment picture, and predictions on economic growth;
- Political perceptions about the health of these industries relative to others;
- Public perceptions about the United State's competitive commercial position, as well as leadership of the United States vs. other nations, in these industries;
- Both real and perceived dependence of U.S. intelligence and national security on leadership in these industries, and U.S. reliance on foreign technologies and assistance in areas related to intelligence and national security; and
- Prevailing political philosophies regarding industrial policy.

---

[4] D. Ernst, 2005, "The New Mobility of Knowledge: Digital Information Systems and Global Flagship Networks," in R. Latham and S. Sassen (eds.), *Digital Formations: IT and New Architectures in the Global Realm*, Princeton University Press, Princeton and Oxford.

[5] D. Ernst, *A New Geography of Knowledge in the Electronics Industry? Asia's Role in Global Innovation Networks*, Policy Studies, No. 54, August 2009, East-West Center, Honolulu, HI, 65 pp.

[6] This is in line with Jacques Gansler's argument for a "global strategy" made for the U.S. defense industry (J. Gansler, 2011, *Democracy's Arsenal: Creating a Twenty-First-Century Defense Industry*, The MIT Press, Cambridge, MA).

The last two decades have exposed significant conflicts among these traditional influences on policy, largely brought about by three important changes: (1) the end of the cold war, (2) the general stagnation of the Japanese economy, and (3) the globalization of the computer and semiconductor industries into well-established[7] and mutually dependent supply chains and markets.

Federal funding of electronic development, from the launch of Sputnik in 1957 almost until the fall of the Berlin Wall, was driven by perceived military requirements, which had significant noneconomic motivation. During this period, significant federal R&D investment was made in innovative semiconductor technology for military application. After some cost reduction and normal technology adoption delay, the same technology and technology roadmap steadily appeared in the commercial market, including advanced compound semiconductors and dramatically new manufacturing equipment, which also found strong commercial adoption, for example, in lithography.

By the mid-1990s this pattern had reversed; that is, the incredible acceleration of the personal computer (PC) and server industries meant that commercial technology was leading rather than lagging behind military technology. This shift led to an increasing focus on the use and adaptation of commercial off-the-shelf technology in federal procurement and contributed to the steady decline in federal funding for R&D,[8] given the U.S. preeminence in the area and the already high levels of research investments by the U.S. computer and semiconductor industry.

Today, cutting-edge R&D in semiconductors, the historical engine of computer performance growth, has become unmanageably expensive for the usual U.S. federal agencies. At the same time, it is extremely difficult for industry to invest in long-term R&D, given the near-term expectations of the financial markets. One consequence has been limited commercial R&D investment in hardware and software technologies whose economic return is not realized rather quickly.

In the United States, industrial policy has typically not been viewed as an offensive tool for economic competition or a means to create new industries or accelerate successful ones, but rather as a defensive tool to protect or restore existing industries under competitive economic pressure.[9] The perception of favoring certain industries, "picking winners," by government pressures and incentives, rather than allowing for natural market forces and laissez-faire investment, has been politically toxic. On the other hand, rescuing at least some foundering industries, or attempting to regain lost ground in critical ones, has been generally politically rewarding. The Asian competitor nations, for example, China and Japan, traditionally have both subsidized and protected (by legal and covert subsidies and tariffs) those industries that they choose to target.

In contrast, it is important to recognize that U.S. industries, and information technology in particular, do not tend to receive attention or assistance from federal sources simply because they are slowing down in growth or maturing; there typically must be a specific adversary. For example, once U.S. superiority in electronics and computation (e.g., for guidance systems) over the Soviet Union became assured, the focus of government policy switched to the rising Japanese dominance in electronics, especially including memories.

While countries such as Japan began forming R&D consortia as early as 1956, the practice was illegal in the United States until Congress passed the National Cooperative Research Act in 1984.[10] Two years later, concerns of a U.S. decline in semiconductor market share prompted a call by the Semiconductor Research Corporation (SRC) and Semiconductor Industry Association (SIA)[11] for increased cooperation to provide the U.S. semiconductor industry with the capability of regaining world leadership in semiconductor manufacturing. As a result of this effort, SEMATECH (Semiconductor Manufacturing Technology) was created in 1987 as a partnership of 14 U.S. semiconductor companies with the Defense Advanced Research Projects Agency (DARPA), which contributed U.S. $500

---

[7]While well established and interdependent, these value chains can be highly vulnerable to sudden disruptions from natural disasters, geopolitical conflicts, and so on. Some of these are discussed in greater detail in Chapter 4.

[8]J. Gansler, 2011, *Democracy's Arsenal: Creating a Twenty-First-Century Defense Industry*, The MIT Press, Cambridge, MA.

[9]U.S. innovation policy can be thought of as "market conforming" in its intent to address problems that economists have deemed weaknesses for technological advancements. In particular, these were externality problems that required collective R&D funding and that funding took specific paths because of appropriation processes in Congress.

[10]D. V. Gibson, and E. M. Rogers, 1994, *R&D Collaborations on Trial*, Harvard Business School Press, Cambridge, MA.

[11]"Founded in 1977 by five microelectronics pioneers, SIA unites over 60 companies that account for 80 percent of the semiconductor production of this country." (see www.sia-online.org) SIA, along with the European Semiconductor Industry Association (ESIA), the Japan Electronics and Information Technology Industries Association (JEITA), the Korea Semiconductor Industry Association (KSIA) and the Taiwan Semiconductor Industry Association (TSIA), sponsors the International Technology Roadmap for Semiconductors, a 15-year assessment of the semiconductor industry's future technology requirements (see www.public.itrs.net). Last accessed on June 30, 2012.

million over 5 years, to solve common manufacturing problems and to regain U.S. competitiveness in the semiconductor industry that had been lost to Japanese industry in the mid-1980s[12]. In the committee's view, SEMATECH played a strong role in early efforts to reclaim U.S. semiconductor manufacturing leadership and has been a successful example of a U.S. consortium demonstrating the value of federal funds and federal participation. This position is reiterated by a 2002 National Research Council report, *Government-Industry Partnerships for the Development of New Technologies*, which found that the SEMATECH partnership directly contributed to the global competitiveness of U.S. industry, specifically the resurgence of the U.S. semiconductor industry.[13]

Today, the SRC also continues to play a significant role in advancing the semiconductor industry though synergetic industry and university research programs and support initiatives around the world, such as the Global Research Collaboration Program, Nanoelectronics Research Initiative, Focus Center Research Program, and Semiconductor Research Corporation Education Alliance. The National Nanofabrication Infrastructure Network (NNIN) also provides a successful example of U.S. government (National Science Foundation) support of university semiconductor research. By paying for some expensive semiconductor research equipment at universities, the NNIN enables leading-edge research, which indirectly supports the U.S. semiconductor industry with research results and science and engineering graduates.

In contrast, two other industry-only consortia, started near the same time and for similar reasons, both failed. The Microelectronics and Computer Technology Consortium (MCC) was formed in the early 1980s as a response to Japan's Fifth Generation Computer Systems (FGCS) project.[14] Entirely funded by corporate partners, MCC worked on a wide range of technology and software projects, with early sponsorship particularly from mainframe computer companies. By 2000 the Board of Directors had decided to dissolve the organization. Another industry-only consortium, U.S. Memories was organized in 1989 to manufacture memories based on technology from IBM, to avoid dependence on Japanese vendors. However, by early 1990 the consortium members had proven unwilling to make the necessary investments, and major memory users like Apple, HP, and Sun did not participate, so the project was canceled. Thus far, consortia that include IT competitors but that do not have government leadership have fared poorly, due to a combination of mutual suspicion, lack of focus, and no real sense of urgency.

In summary, U.S. federal support and investment has historically relied upon a perception of military threat, economic decline, industry crisis, and/or loss of competitive position; and in the United States, electronic and computer consortia without both federal R&D support and federal direction have not generally succeeded. Thus, centralist technology policies that may work in nations and cultures that accept such direction readily are a poor match to the U.S. free-market model. Further, innovation policy has to reflect each country's unique economic institutions, industry structure, and growth model.

### 3.1.2 Global Semiconductor Competition

While it could be proposed that some U.S. computer vendors "failed to innovate," or "gave up the fight" to foreign competition, it is important to recognize the paired advantages and shortcomings of a free-market industrial economy, and the capacity it provides for innovation, not only in technology, but in the creation (and destruction) of whole economic sectors. U.S. capital market investors are often quick to spot and to capitalize on transformative shifts in a business paradigm, and, consequently, to move their investments in a way that often accelerates the change. Capital markets tend to value short-term quarterly profits and tend to reward or punish a company accordingly, which manifests in changes in its stock price. This has advantages and disadvantages. On the one hand, it discourages waste and encourages competitiveness. On the other hand, a short-term focus often discourages long-term thinking and R&D investment, particularly during difficult economic times. Federal R&D programs and public-private consortia play a crucial role in coping with this tension.

In the late 1980s, when it appeared that focused government programs in Japan, as well as unfair or unreasonable trade practices, might overtake U.S. competitiveness, DARPA investments, especially SEMATECH, drove the necessary R&D efforts in process and equipment to sustain Moore's Law and to maintain the confidence of capital markets. Concurrently, IBM began to accelerate its investment in very high-performance semiconductor technology and to form joint innovation partnerships with numerous (non-

---

[12] See www.sematech.org/corporate history; www.sematech.org/corporate/timeline; NRC, 2003, *Securing the Future: Support to the Semiconductor Industry*, Washington, D.C.: The National Academies Press (available online at http://www.nap.edu/catalog.php?record_id=10677#toc).

[13] NRC, 2002, *Government-Industry Partnerships for the Development of New Technologies*, Washington, D.C.: The National Academies Press (available online at http://www.nap.edu/catalog.php?record_id=10584).

[14] Kazuhiro Fuchi, 1984, *Revisiting Original Philosophy of Fifth Generation Computer Systems Project*, FGCS, pp. 1-2.

Intel) semiconductor fabrication companies, creating a business counterpoint to Intel. More recently, the rise of mobile computing devices has created new competition, both among existing companies and new ones formed in response to emerging market economies.

### 3.1.3 Creation of the U.S. (and Global) Software Industry[15]

Continuous technical innovation that sustained Moore's Law (and exponential growth in computing performance) led also to the creation of the commercial software industry as a meaningful force in the U.S. economy. This took place in two ways. First, the falling cost and wide availability of powerful microprocessors greatly increased the number of computers in use, and successful software products could be sold in enormous numbers at modest prices.

Second, the fact that a small number of instruction set architectures (ISAs) dominated the PC and server marketplace[16] meant there was a larger and consolidated software market that would benefit from steady improvements in cost and performance, while seldom requiring any significant changes to the programs. Vendors rarely prefer to use new instructions sets until they have been in the market for many years and are available on a significant fraction of deployed machines. This allowed larger software investments to be made, in products that would surely perform better over time, courtesy of Moore's Law.

U.S. firms dominate this 30-year-old PC and server industry, although a few European firms (e.g., SAP) are of significant size and share of the market. There is early evidence that this market dominance may extend to the new world of smartphones and tablets, as well as cloud services, though global competition in this space is new and intense.

### 3.1.4 Consequences of the U.S. Free-Market Approach

These three phenomena—the enormous growth of the semiconductor industry, the commoditization of the computer industry, and the emergence of a huge software industry are mutually dependent, and have created a virtuous economic framework. They also afforded the United States the opportunity to achieve and maintain its leadership in information technology generally. Although federal support to long-term R&D has been indispensible, particularly in bad economic environments, such achievements would almost certainly not have been possible under a centrally managed policy regime. For instance, in a more managed environment, the policy impulse might have been to save legacy computer companies; instead, market forces coupled with a noninterventionist approach and (to a lesser extent) government antitrust efforts helped ensure the "creative destruction" that has transformed computing and the role of the United States in it. The United States has been rewarded by the emergence of very strong semiconductor design and software industry leadership—in exchange for the loss of some semiconductor fabrication and the assembly and testing of commodity products to foreign vendors, for example, the off-shore assembly by contract manufacturers of even the strongest U.S. computer brands, based upon cost.

However, staying strictly on any technical path involves bypassing others, and sacrificing progress in some areas to sustain others. It is certainly worth examining some of the approaches delayed or abandoned by the course taken by the IT industry.

In the 1960s and 1970s, computers were expensive, resources were limited, and programmers were scarce. There was great emphasis on creating clever algorithms that required the least number of instructions or smallest amount of memory, or both. Elegant, parsimonious program design was celebrated, and improvements to compilers for denser code and new languages for programmer productivity were high priorities in academia and industry alike. High-productivity programming languages help programmers produce working programmers faster, as compared with high-performance programming languages, which help programmers extract as much performance as possible by exposing machine details to them. From high-productivity languages and the relentless hardware performance improvements enabled by Moore's Law, a new and much larger pool of programmers emerged. These programmers applied application-specific knowledge, for example, machine learning, graphics, animation, accounting, government functions, and so on, driving an explosion in software capabilities in the era of ever-faster and cheaper central processing units (CPUs) and memory. In addition, programming emphasis moved from performance productivity to getting new products out faster.

Another area of technology research and innovation affected by Moore's Law was parallel computing. In the 1970s, very large scientific computers with parallelism among several arithmetic units were just beginning to

---

[15]Section 3.1.3 and part of Section 3.1.4 rely heavily on David Liddle, "The Wider Impact of Moore's Law," *IEEE SSCS Newsletter*, September 2006. Available at http://ieeexplore.ieee.org/stamp/stamp.jsp?arnumber=4785858. Last accessed August 21, 2012.

[16]The history of the mobile and embedded computing space is much more varied, with a diversity of ISAs and vendors.

work well. Equally important, software researchers were beginning to make real progress on the problem of programming systems for parallel machines. Moore's Law advances relegated most work in parallel computing to business servers and scientific and technical computing.[17] A revival of parallel computing research and development in the 1990s yielded several new approaches and companies, but the early promise was not realized, for many reasons. Because the size of the technical computing market was small relative to the PC market, the research and product development took a different path, focusing on performance maximization at reduced costs rather than ease of use and programmer productivity.

Only now that the limits of growth in CPU clock frequency are in sight for consumer devices has serious focus on parallel processing reentered the mainstream. Earlier research from the 1980s and 1990s showed the difficulty of developing tools that can easily convert legacy sequential codes into scalable parallel code that run well on current generation or next-generation machines. This research experience suggests that future work should emphasize simplicity and programmability of heterogeneous multicore devices to address mainstream product needs. In addition, such research should be driven by an awareness of the demonstrated limitations of automatic parallelization and recognition that the intrinsic parallelism in application problems differs markedly.

Another effect of Moore's Law over the years has been that the capital markets, given the visible, vast investment in scalable complementary-symmetry metal-oxide-semiconductors (CMOS) (except for extremely low-volume exotic noncommercial uses) have not encouraged R&D on new post-silicon materials. Even small deviations from the CMOS path, like silicon-germanium or silicon carbide have aroused skepticism, let alone compound III–V semiconductors.

Typically, a free-market approach will continue on a profitable path until it begins to reach diminishing returns; unfortunately, that point is sometimes recognized too late. The uniprocessor CMOS clock-frequency race has already ended, leaving the United States ill prepared with either semiconductor or software succession plans. This type of situation has traditionally been one in which U.S. government participation with academia and industry has been effective. Neither the U.S. federal government nor the U.S. computer industry has come to consensus on a strategy that ensures U.S. leadership in the next generation of computing technologies. Thus, those developing future U.S. policy in these areas should carefully consider the opportunities and consequences of alternative innovation strategies, some of which have been tried elsewhere.[18]

## 3.2 China – Strengthening Indigenous Innovation

Over the last several decades, China has made significant efforts to align its science, technology, and innovation policies to support indigenous innovation. These trends towards technonationalism[19] were prompted by political concerns within China that it both lacks indigenous technology and depends on foreign technology, as well as from several lessons learned over the past several decades. For example, after its relationship with the former Soviet Union ruptured in the 1950s, China shed its reliance on Soviet technology and developed a national strategic weapons program, developing its own nuclear weapons, missiles, and satellites.

Then, in the late 1970s, China embraced globalization by opening its huge market to multinational corporations for the exchange of advanced technology. In turn, the rapid growth of China's semiconductor consumption primarily reflects its emergence as the dominant global factory for IT equipment. Between 2004 and 2009, alone, China's share of global electronic equipment production increased from 17 percent to 31

---

[17]It is worth noting that parallel computing work continued in the high-performance computing (HPC) and server segments.

[18]When examining innovation strategies elsewhere, it is important to recognize that most are being applied to those whose attempts to enter the advanced computing market were late relative to the United States. Latecomers have disadvantages and advantages that need to be assessed and taken into account when considering policy options and what lessons might be learned. For instance, latecomers to advanced computing need to overcome very substantial entry barriers (disadvantages), as well as to exploit new opportunities that result from beginning with less-complicated technology and having fewer legacy constraints on technology development, strategy, and organization (advantages). Economies of scale may be a barrier to market entry requiring nonprice means of market penetration, that is, through product differentiation and the creation of new markets and distribution channels. On the other hand, latecomers who become fast followers of established technology roadmaps are able to set clear targets for product development and related research, as well as to compare and learn from the experience and failures of incumbent leaders. Latecomers are also not locked into supporting and maintaining legacy technologies or infrastructures.

[19]Policy orientation towards autonomy and independence from other states (see B. Naughton and A. Segal, 2001, "Technology Development in the New Millenium [sic]: China in Search of a Workable Model," *MIT Japan Program*, Working Paper Series 01.03., May 28).

percent.[20] This marked increase suggests a reshaping of the IT equipment manufacturing landscape.

However, as the global financial and economic crisis continues, exports from China have slowed, placing pressure on China's export-oriented economic development model. In pursuit of new growth engines for its economy, several government policies and initiatives have facilitated the strengthening Chinese indigenous innovation.

### 3.2.1 Government Policies and Initiatives to Strengthen Indigenous Innovation

*Medium and Long-Term Plan (2006–2020)*

In early 2006, China released its Medium- and Long-Term Plan for the Development of Science and Technology (MLP) (2006–2020). This plan set the tone for strengthening China's indigenous innovation capability by addressing four problems in China's scientific and technological development: (1) lack of innovation in commercial technologies and dependence on foreign technology; (2) increasingly unfriendly international environment for acquisition of foreign technologies; (3) technological failure to meet critical energy, water and resource utilization generally, and environmental protection and public health needs; and (4) mounting technological challenges for meeting national defense needs.[21] While not specifying what indigenous innovation means, the MLP highlights three channels through which indigenous innovation capabilities may be strengthened: (1) genuine "original innovation," (2) "integrated innovation" (fusing together existing technologies in new ways), and (3) "reinnovation" (assimilation and improvement of imported technologies).

The MLP singles out 16 engineering mega-programs, as well as identifies 11 key areas, 8 frontier technologies, and 4 science mega-programs, to support in the next 15 years (see Appendix I). Many of these programs and focus areas are directly relevant to advanced computing, including the IT industry and modern services (key areas); information technology and new and advanced materials (frontier technologies); and core electronic components, high-end generic chips, basic software, extra-large-scale integrated circuit (IC) manufacturing and techniques, and new-generation broadband wireless mobile telecommunications (mega-engineering programs).[22]

*Strategic Emerging Industries (SEI) Program*

Strategic emerging industries (SEIs) refer to industries[23] associated with the development of technologies (e.g., information, biotechnology, medical, new energy, environment, marine, and space) that have strategic importance to China; many are similar to the frontier technologies prioritized in the MLP. These SEIs have been said to represent the future direction of industrial development in China and will play a critical role in its continuous and sustainable economic growth, particularly in national economic and social development and optimization and upgrading of industrial structure.

Launched in October 2010,[24] the SEI Program was highlighted as an important component of the 12th Five-Year Plan for National Economic and Social Development (2011–2015). As selected SEIs are science and technology based, the SEI Program is expected to decrease China's dependency upon external technology and boost indigenous innovation capabilities, ultimately spurring economic growth and the formation of a new industrial cycle. It is expected that the government will also work out financial and taxation policies to support, guide, and encourage capital investment, and establish special funds for the development of SEIs.

Foreign-invested design subsidiaries of leading foreign semiconductor companies and global original equipment manufacturers (OEMs) play an important role in China's chip design industry. Of the 472 design enterprises reported in China at the end of 2009, approximately 100 were the design units or activities of foreign-invested or subsidiary multinational

---

[20] PwC, 2011, *Continued growth: China's impact on the semiconductor industry – 2011 update*, p 13. Available at www.pwc.com/gx/en/technology/assets/china-semiconductor-report-2011.pdf. Last accessed January 27, 2012.

[21] C. Cao, R. P. Suttmeier, D. F. Simon, 2009, "Success in State Directed Innovation? Perspectives on China's Plan for the Development of Science and Technology," in G. Parayil and A. P. D'Costa (eds.), *The New Asian Innovation Dynamics: China and India in Perspective*, Palgrave Macmillan, London, pp. 247–264.

[22] See Alan Wm. Wolff, "China's Drive Toward Innovation," *Issues Online*, Spring 2007. Available at http://www.issues.org/23.3/index.html. Last accessed August 21, 2012.

[23] SEIs include the following: Energy-saving and environmental protection, new generation of IT, biotechnology, and high-end equipment manufacturing industries; new energy, new materials, and new energy automobile industries. State Council of China, 2010, *Decisions of State Council on Accelerating the Cultivation and Development of Emerging Strategic Industries*, G.F. No.32, October 29 [USITO Draft Translation].

[24] On October 18, 2010, the State Council issued a "Decision on the Acceleration of Nurturing and Developing Strategic Emerging Industries," formally launching the SEIs. The 12th Five-Year Plan for National Economic and Social Development (2011–2015), released in 2011, includes SEIs as one of its important components.

companies.[25] These foreign-invested design subsidiaries engage in a variety of activities that range from the simple to the complex, including adapting parent company product standards for the China market, providing lower-cost capacity for standardized back-end design functions that are integrated in the parent company's design flow, but they also include integrated design projects for system-on-a-chip (SoC) designs.

In addition to multinational partnerships, long-term investments in the IT and advanced computing industries are now beginning to yield competitive, indigenous microprocessors, designed wholly in China. Over the last decade, the Chinese Academy of Sciences (CAS) has been developing and prototyping their line of Loongson and Godson processors.[26,27] During this time, each generation of Loongson or Godson processor has become more capable and is now rivaling leading-edge microprocessors in performance. Similarly, the ShenWei series of microprocessors,[28] developed by the Jinan Institute of Computing Technology (affiliated with the People's Liberation Army) since 2006, has found its way into the first home-grown supercomputer in China, called BlueLight. As of November 2011, BlueLight includes 8,704 ShenWei chips and achieves Linpack performance of 795 TFlops,[29] while being one of the most energy-efficient computers in the world.

While neither of these programs has shown any commercial success either inside or outside of China, the investment period is long and ongoing. One would expect that such programs aspire to serve the Chinese domestic market, at least for high-end technical computing. As these, and other, systems diverge from the x86 and ARM ecosystems, they will have different sets of innovation and performance optimization challenges. It is likely that national innovation policies, both in China and in the United States, will have to be iteratively redefined to meet these challenges. This is also the case should Chinese-designed computing systems begin to have any commercial success inside or outside of China.

It is worth noting that China does not have any representation among the largest semiconductor companies, nor does it have any mass commercial processors or architectures. China does, however, have a seat at the six-seat World Semiconductor Council. This seat may reflect both China's growing potential as a semiconductor market and its increasing capabilities in semiconductor design. China's increasing share of worldwide patents focused on semiconductor technology, from 13 percent to 22 percent in 2009, also suggests improvements in China's semiconductor innovative capacity. More significantly, China's share of semiconductor patents that are first issued in China has grown from 0 percent in 2005 to ~24 percent in 2009.

### 3.2.2 Impacts of Government Policy Efforts

It is unclear to what degree the MLP will enhance China's indigenous innovation capability. However, international technical and business communities have already expressed concern over Chinese efforts to support the indigenous innovation efforts. For example, government policies, which gave preference for procuring domestic technologies and products in the name of supporting indigenous innovation, were abolished, at least temporarily, due to extreme pressure from foreign governments and companies. In addition, some studies[30] have characterized China's innovation policies as a threat to global intellectual property rights; a recent report by the U.S. Chamber of Commerce has even claimed that Chinese innovation policy is "a blueprint for technology theft on a scale the world has never seen before."[31]

Lastly, the committee believes, growth in China's homegrown industrial capacity, plus China's massive urbanization, has nurtured an increasingly large domestic market in different manufacturing—and increasingly R&D—sectors. It is not clear to what degree other key

---

[25]PwC, 2011, *Continued growth: China's impact on the semiconductor industry – 2011 update.* Available at www.pwc.com/gx/en/technology/assets/china-semiconductor-report-2011.pdf. Last accessed January 27, 2012.

[26]Godson 3B is an 8-core processor with vector extensions implemented in a 65 nm technology. It employs a MIPS instruction set (originally developed in the United States), runs at 1 GHz, and has a peak performance of 128 GF.

[27]The Loongson and Godson processors were developed under the leadership of U.S.-trained computer scientist Li Guojie at the Chinese Academy of Sciences' Institute of Computing Technology, CAS.

[28]The ShenWei 3 chip contains 16 cores, is implemented in 65nm and runs at 1.1GHz.

[29]There are two other Chinese HPC installations above BlueLight (no. 14) in the November 2011 TOP500 List. However, both of these are built from Intel CPUs and NVIDIA graphics-processing units, delivering 2.566 PFlops (no. 2) and 1.271 PFlops (no. 4). The BlueLight cluster is ranked no. 39 on the Green500 List of the most efficient supercomputer clusters.

[30]U.S. International Trade Commission, 2010, *China: Intellectual Property Infringement, Indigenous Innovation Policies, and Frameworks for Measuring the Effects on the U.S. Economy*, November.

[31]James McGregor, 2010, *China's Drive for "Indigenous Innovation": A Web of Industrial Policies*, Global Intellectual Property Center and Global Regulatory Cooperation Project under the U.S. Chamber of Commerce, and APCO Worldwide, Washington, D.C. Available at http://www.apcoworldwide.com/content/PDFs/Chinas_Drive_for_Indigenous_Innovation.pdf. Last accessed on August 7, 2011.

industries will follow similar trends, in particular, those SEIs related to the advanced computing technologies and products. In overall growth, the added value of SEIs has been estimated to reach U.S. ~$682 billion in 2015 and U.S. ~$1.8 trillion in 2020, with projected annual growth rates of 24.1 percent between 2011 and 2015 and 21.3 percent between 2016 and 2020.[32]

*3.2.3 Transition Toward Economic Outcomes-driven S&T Programs*

Unlike most previous government-led science and technology (S&T) programs, where the Ministry of Science and Technology (MOST) was the only or biggest stakeholder, SEI program development efforts have primarily been led by the National Development and Reform Commission and Ministry of Industry and Information Technology (MIIT), two superministries with strong economic missions in China's bureaucracy.[33] In contrast to MOST, these agencies are expected to increase industry participation in the program (as opposed to primary participation by universities and research institutes). As such, it is likely that future S&T programs will require strong economic components and targets and cannot be assessed by publications and patents alone. The fact that economic instead of science agencies are deeply involved in the SEI Program reveals a long-standing issue—the separation between research and the economy—that China has tried to solve when it started to reform its S&T management system in the mid-1980s. Regardless, the successful transformation of the Chinese economy will depend upon the successful coordination among different government ministries, each with a unique mission. Achieving this will be difficult, if not impossible.

**3.3 Taiwan – Low-cost and Fast Innovation**

In the last several decades, Taiwan has experienced incredible economic growth—particularly in its IT industry, which now accounts for 70 percent of Taiwan's total manufacturing R&D.[34] To date, a defining characteristic of Taiwan's IT industry has been its deep integration into diverse and global corporate networks of production and innovation. In addition to facilitating the role of Taiwanese firms as fast followers, this network integration also encouraged IT firms to focus their R&D efforts on incremental innovation. Today, there is a growing recognition that Taiwanese firms must now increase and broaden R&D in order to avoid diminishing returns of network integration. Thus, new policies are needed to spur domestic capabilities for low-cost innovation in IT.

While Taiwan's new innovation strategy is still a "work in progress," some major policy building blocks, which combine market-led innovation and public policy coordination of multiple layers of private and public innovation stakeholders, are taking shape. Due to its pragmatism and openness to new forms of public policy and private-public partnerships, Taiwan's innovation policy may in fact shed new light on the opportunities and challenges for strengthening America's innovation capabilities in advanced computing.

*3.3.1 Taiwan's "Global Factory" Innovation Model*[35]

Early on, Taiwan's IT industry depended heavily on international markets and access to foreign technology, tools, and ideas to overcome substantial entry barriers to network participation, namely, a lack of domestic market and limited resources and infrastructure. From the beginning, the key to Taiwan's success has been an early integration into diverse and constantly evolving network arrangements that include both formal corporate and informal knowledge networks. Formal corporate production networks link Taiwanese firms to large global brand leaders (the customers), investors, technology suppliers, and strategic partners through foreign direct investment as well as through venture capital, private equity investment, and contract-based alliances. Equally important are informal global knowledge networks that link Taiwan to more developed overseas innovation systems and knowledge communities, primarily in the United States, through the international circulation of students and knowledge workers.[36] Finally, domestic interorganizational linkages with large Taiwanese

---

[32]Zhou Zhizue, chief economist on MIIT projections. Available at http://tech.sina.com.cn/it/2011-08-04/07185880063.shtml. Last accessed on November 7, 2011.

[33]Barry Naughton, 2009, "China's Emergence from Economic Crisis," *China Leadership Monitor*, No. 29. Available at http://media.hoover.org/sites/default/files/documents/CLM29BN.pdf. Last accessed on November 7, 2011.

[34]Between 2001 and 2006, almost 90 percent of the R&D investment of Taiwan's private sector was concentrated on two sectors, electronics components (56 percent) and computers and electronic and optoelectronic products (32 percent). See http://eng.stat.gov.tw/ct.asp?xItem=6503&CtNode=2202&mp=5.

[35]Sections 2.3.1- 2.3.4 rely heavily on Dieter Ernst, "Upgrading through innovation in a small network economy: insights from Taiwan's IT industry", *Economics of Innovation and New Technology*, June, 2010 Volume 19, No. 4pp. 295-324. Last accessed August 21, 2012.

[36]Between 1987 and 2003, this small island has been the fifth largest nation of origin of international students in the United States. (Guo, 2005: 142).

business groups complement these international linkages.[37]

Progressive integration into these diverse production, knowledge, and innovation networks has enabled Taiwanese firms to combine the speed and flexibility of smaller firms with the advantages of scale and scope that normally only large firms can reap, as well as to tap into the world's leading markets, especially in the United States. Network participation has also multiplied conduits for knowledge transfers to Taiwanese IT firms, broadening their scope for learning and capability development. This, in turn, has created new opportunities, pressures, and incentives for Taiwanese IT firms to upgrade their technological and management capabilities and the skill levels of workers.

Today, Taiwan has established itself as an important "global high-tech factory" for PC-related products, handsets, wireless equipment, integrated circuits, and flat panel displays.[38] For global IT industry leaders, Taiwanese firms have become preferred OEM and ODM (original design manufacturing) suppliers.[39] In addition, Taiwanese firms have made considerable progress in product development, especially in electronic design. Beginning in the 1980s, Taiwan's leading PC firms established R&D laboratories in Silicon Valley to gain early access to the product and technology road maps of the global industry leaders and to improve their product-development capabilities. By the mid-1980s, Taiwan's semiconductor firms became involved in board-level and application-specific integrated circuit design,[40] giving rise to a broad portfolio of design implementation capabilities. This enabled Taiwanese semiconductor firms to compete on the speed, cost, flexibility, and quality of providing these services.

However, over the last decade, the globalization of S&T and of the economy has placed pressures on Taiwan's IT industry. Investments necessary for radical innovations that can compete with global technology leaders are beyond the reach of many Taiwanese IT companies. Even TSMC (Taiwan Semiconductor Manufacturing Company), the world's leading IC foundry, has had to stretch its resources to the limit to sustain its leadership position. In addition, Taiwanese IT firms are establishing low-cost supply bases in China and Southeast Asia to reduce production costs. To expand their position as network suppliers, Taiwanese firms are also moving beyond the provision of manufacturing services, and developing integrated service packages that include logistics and product development. However, the downturn in the global electronics industry since late 2000 has exposed several challenges for Taiwan's innovation model.

*3.3.2 Negative Effects of Network Integration*

Taiwan's focus on the provision of OEM and ODM services has severely constrained the capacity of Taiwanese firms to invest in "upgrading through low-cost innovation" strategies.[41] This problem is exacerbated by relentless pressure from global brand marketers to reduce cost and time-to-market for commodity-type products with low profit margins that are apt to penetrate mass markets. As a result, Taiwanese firms are stuck in a "commodity price trap," with insufficient profit margins to support investment in new R&D, intellectual property creation and branding. Furthermore, as specialized OEM and ODM suppliers, Taiwanese firms typically concentrate on incremental innovations within existing product architectures that are predefined by global brand leaders charging hefty patent-licensing fees. This structure constrains their capacity to develop new products and to shape technology road maps and standards.

Since 2003, many Taiwanese handset makers have attempted to increase profits by increasing their branded handset sales relative to their OEM or ODM business. However, with the possible exception of ASUS and HTC,[42] most of these attempts seem to have failed, causing Taiwanese handset makers to switch back to the OEM-ODM model. The most spectacular failure has

---

[37]D. Ernst, 2001, "Small Firms Competing in Globalized High Tech Industries: The Co-Evolution of Domestic and International Knowledge Linkages in Taiwan's Computer Industry," in P. Guerrieri, S. Lammarino, and C. Pietrobelli (eds.), *The Global Challenge to Industrial Districts: Small and Medium-Sized Enterprises in Italy and Taiwan*, Edward Elgar, Aldershot, U.K.

[38]D. Ernst, *Upgrading through innovation in a small network economy: insights from Taiwan's IT industry*, Economics of Innovation and New Technology, Vol. 19, No. 4, June 2010, pp. 295–324.

[39]An OEM contract refers to arrangements between a brand-name company (the customer) and the contractor (the supplier), where the customer provides detailed technical blueprints and most of the components to allow the contractors to produce according to specifications. In ODM arrangements, the contractor is responsible for design and most of the component procurement, with the brand-name company retaining exclusive control over marketing.

[40]D. Ernst and D. O'Connor, 1992, *Competing in the Electronics Industry. The Experience of Newly Industrialising Economies*, Development Centre Studies, OECD, Paris, 303 pp.

[41]D. Ernst, *Upgrading through innovation in a small network economy: insights from Taiwan's IT industry*, Economics of Innovation and New Technology, Vol. 19, No. 4, June 2010, pp. 295–324.

[42]HTC has successfully developed own-brand touch-screen smartphones, initially based on Microsoft's Windows Mobile operating system, but now also on Google's open-source Android platform.

been the attempt by the BenQ Group (a spin-off of the Acer Group) to accelerate its global branding strategy by acquiring the mobile handset business of Siemens and its intellectual property.[43]

### 3.3.3 Constraints to Developing Indigenous Intellectual Property

While Taiwan's patent filings at the U.S. Patent and Trademark Office (USPTO) have grown rapidly (in "all patents" per million of its population and in "utility patents"[44]), its patent counts are highly concentrated, both in terms of products (technology classes) and patent holders (assignees). In 2010 the largest number of Taiwan's U.S. patents[45] were in semiconductor manufacturing, and these patents were dominated by TSMC (Taiwan's third-largest patent filer, with 405 patents), followed by MediaTek (no. 5), Macronix (no. 6), United Microelectronics Corporation (no. 7), and Via Technologies (no. 9, with 108 patents). Hon Hai (Taiwan's largest patent filer, with 572 patents), the world's largest maker of electronic components, has pursued an aggressive strategy to file protective patents (especially for its connector technology), primarily against China.[46] In addition, Taiwan's patent quality remains low (in patent citation, "science linkages," and technological capabilities),[47] and its most influential patents are highly concentrated with TSMC.

Taiwan's IC design industry provides a telling example of the substantial challenges of developing indigenous intellectual property. As specialized suppliers to global semiconductor and system companies, Taiwanese chip design firms have limited resources and incentives to close the technology gap relative to industry leaders—and as a result, they are typically not active at the leading edge of process technology and IC complexity.[48] In addition, Taiwanese design houses have not been able to develop in-house complete solution packages. For instance, in the important cellular chipset market, only one Taiwanese design house (MediaTek) offers a complete cellular chipset solution. All other Taiwanese companies competing in this market, such as Sunplus and Airoha, have focused on specific building blocks and niche markets. Given the rapid change and unpredictability of these markets, such a focused approach is a high-risk strategy.

### 3.3.4 Hollowing-out Through Offshoring to China

To retain its position as OEM and ODM suppliers to global brand marketers, Taiwan has established low-cost supply bases—and more recently, R&D centers—in China and Southeast Asia. The increasingly common practice of Taiwanese IT manufacturers receiving orders in Taiwan and shipping manufactured goods from China has given rise to "a new cross-Strait division of labor along the lines of pilot run vs. mass production."; however, this offshore outsourcing now imposes severe pressures on Taiwan's IT industry, as reflected by a declining domestic value-added ratio that is much lower than for the United States and Japan.[49,50]

Another important concern is the continuing relocation of wafer fabrication capacity. Although China's current wafer fabrication capacity represents about 9.4 percent of worldwide wafer fabrication capability,[51] as of May 2011, 28 new wafer fabrication facilities were under construction in Greater China. Once these begin production, it is estimated that China and

---

[43]Less than 1 year after the acquisition, the German subsidiary, BenQ MobileGmbH & Co OHG, was closed amid continuing huge losses at the subsidiary. BenQ's share of the Taiwan handset market now languishes around 8 percent. BenQ now outsources handset production to Taiwanese contract manufacturers.

[44]A utility patent protects any new invention or functional improvements on existing inventions (such as going from light-emitting diode [LED] technology to organic LED technology), while a design patent protects the ornamental design, configuration, improved decorative appearance, or shape of an invention. China's *utility model patents* protect any new technical solution relating to the shape and/or structure of a product, which is fit for practical use. Utility patents, which offer the same protection (albeit for a shorter time span) as invention patents, are quicker and cheaper to obtain, because they only receive a preliminary examination rather than the full substantive examination of an invention application.

[45]See http://www.uspto.gov/web/offices/ac/ido/oeip/taf/asgstc/twx_ror.htm. Last accessed on January 12, 2012.

[46]Hon Hai has been expanding its USPTO patent portfolio, more than doubling its USPTO patent filings between 2006 and 2010. Since 1995, 61 percent of Hon Hai's patents were filed in China, against less than 18 percent in the United States.

[47]Xin-Wu Lin, 2005, *An Analysis of Taiwan's Technological Innovation – On the Basis of USPTO Patent Data Analysis*, slide presentation, Taiwan Institute of Economic Research, Taipei, July 27. For instance, Taiwan's patents are less "original" and have less "impact" than Korea's, that is, they are less frequently cited within a technology class and in other technology classes. As for science linkages, Taiwan's patents, even for semiconductors, are less frequently cited in scientific journals than Korea's patents.

[48]Joy Teng, 2006, IC Design House Survey 2006: Taiwan, courtesy of *Electronic Engineering Times Taiwan* (available at www.eettaiwan.com).

[49]This hollowing-out effect, and the resultant job displacements, may have been reduced by the growth of Taiwanese exports to Asia (especially China) of increasingly sophisticated production equipment.

[50]S.-H. Chen, M.-C. Liu, and K.-H. Lin, 2005, "Industrial Development Models and Economic Outputs: A Reflection on the 'High Tech, High Value-Added' Proposition", manuscript, Chung-Hua Institution for Economic Research, Taipei: p. 25.

[51]SEMI Wafer Fab Watch, May 2010.

Taiwan together will have a 29 percent share of total worldwide wafer fabrication capacity.

As Taiwanese offshoring extends beyond manufacturing into product development, the competitive advantages previously afforded by Taiwan's high-tech cluster (i.e., combination of flexibility, low cost, and timely service) have begun to erode. For example, as production of computer, communications, and consumer products moves to China, Taiwan's IC design houses are forced to follow suit to sustain close interaction with their customers. Once in China, Taiwanese design houses face intense competition from lower-cost Chinese competitors, and they lose their most fundamental competitive advantage: access to a pool of highly trained and experienced lower-cost engineers and managers from diverse sources. Even worse, Chinese IC design firms can now draw on Chinese returnees who have studied and worked in the United States, as well as recruit former employees of Taiwanese companies to train China's growing pool of local engineering graduates.

### 3.3.5 Government Policies to Support Low-Cost and Fast Innovation

There is a growing consensus in Taiwan that an exclusive focus on hardware manufacturing is no longer sufficient to guarantee sustainable growth. Taiwan's new innovation strategies now seek to build on its capacity for low-cost and fast manufacturing by complementing its contract manufacturing and component production excellence with knowledge-intensive support services and a capacity to provide "integrated solutions." In addition, Taiwan has a long-term objective to strengthen its software capabilities, especially for the design of complex system software and for cloud-computing applications.[52] To implement this strategy, Taiwan's innovation policies seek to strengthen further the linkages and interactions among industry, academia, and public and private R&D organizations.

A defining characteristic of Taiwan's innovation policy is its openness to foreign strategic advice and knowledge sharing, distinguishing it from Japan, Korea, and China[53] with their much more closed systems of innovation policy.

In addition to providing aggressive tax incentives,[54] Taiwan's innovation policy seeks to strengthen the lead role of the private sector by generating new public-private partnerships and by coordinating their interactions.[55] In particular, government initiatives, such as Taiwan's Technology Development Programs, Hsinchu Science Park, and Industrial Technology Research Institute (ITRI), are intended to foster industrial upgrading through low-cost and fast innovation. Today, Hsinchu Science Park is the world's leading cluster for semiconductor manufacturing. ITRI also continues to play a significant role in Taiwan's IT and semiconductor industries. ITRI's recent Cloud Computing Center for Mobile Application (CCCMA) seeks to promote Internet-based, on-demand computing (cloud computing) as a catalyst for strengthening Taiwan's software capabilities, building on Taiwan's strengths in lower-cost hardware, such as memory, chipsets, server, and storage network equipment.

### 3.3.6 U.S.-Taiwan-China Linkages

Since its inception, Taiwan's IT industry has greatly benefited from its deep integration with America's innovation system, especially Silicon Valley. As a byproduct, the United States and Taiwan have developed a strong mutual dependence on each other's IT and semiconductor industries. U.S. IT companies remain the most important buyers of Taiwanese ODM and OEM services, and Taiwan's silicon foundries are a critical supplier of process technology as well as manufacturing and design services to U.S. fabless design companies. In addition, Taiwan exploits a first-tier supplier advantage due to the establishment of leading U.S. R&D centers in Taiwan and to the acceleration of its "upgrading through innovation" strategy.

However, these relationships have been complicated by the emergence of China as both a partner and competitor with Taiwan. In the last decade, China has become not only the most important production site for Taiwan's IT companies, but also a major growth market. Not only are Taiwan's foundries, IC design houses, and ODM suppliers well placed to exploit China's rapid-

---

[52] Interview with Dr. Tzi-cker Chiueh, General Director, ITRI-CCCMA, April 25, 2011. As is typical for Taiwan's leading innovation actors, Dr. Chiueh's education and employment history shows strong links with the United States. See also Ministry of Economic Affairs (MOEA), 2011, Taiwan's ICT industry development and outlook, as reported in *DigiTimes*, August 29.

[53] While China has, to some degree, followed the Taiwanese low-cost and fast innovation model, the Chinese model differs in that it has not leveraged domestic and, to a lesser extent, global innovation networks (see *Run of the Red Queen: Government, Innovation, Globalization, and Economic Growth in China* by D. Breznitz and M. Murphree, 2001).

[54] Taiwan's Statute for Industrial Innovation has lowered the business tax from 25 percent to 17 percent in 2010 (which compares to China's 25 percent rate, Korea's 22 percent rate, and Singapore's 17 percent rate).

[55] H.-S. Chu, 2007, "The Taiwanese Model: Cooperation and Growth," in C. W. Wessner, ed., *Innovation Policies for the 21st Century. Report of a Symposium*, The National Academies Press, Washington, D.C., p. 120.

demand growth for IT products and services, but Taiwan's government is convinced that China is gradually becoming a regional technology leader. This reliance has resulted in new initiatives for cross-strait cooperation in industrial standards, for broader bilateral economic cooperation, especially through the Economic Cooperation Framework Agreement (ECFA),[56] and for deregulation of Chinese investment in Taiwan. On the other hand, continuous penetration of the Chinese market will require that Taiwanese firms also redeploy new product development and research to China. By providing critical inputs (through training, technology transfer, and joint product development) to Chinese firms, Taiwan accelerates China's ability to catch up.

As Taiwan's IT industry becomes increasingly integrated with China's economy and its innovation system, it is unclear how and to what degree Taiwan will strike a balance between cooperation with China and cooperation with the United States. If the sheer weight of China forces Taiwanese firms to give priority to their links with China, how will this affect America's access to the semiconductor global value chain? It is too early for a conclusive answer to these questions. So far, however, Taiwan's economic diplomacy related to the IT industry remains closely aligned with the U.S. position.[57]

*3.3.7 Summary*

If Taiwan is to survive intensifying technology-based global competition, it must move beyond its traditional "global factory" innovation model, which will require quick access to radical innovations, especially in generic technologies. While Taiwan has significant policy initiatives in each of the above areas,[58] the risk of failure remains high, implying that an exclusive focus on technology leadership strategies is unlikely to support a broad-based upgrading through innovation strategy. These risks explain why Taiwan's new innovation strategy emphasizes low-cost and fast innovation through domestic and global innovation networks. Recent policies suggest that China is following suit with Taiwan's innovation model and will focus in the future on low-cost mass adoption of new technologies and innovation.

## 3.4 Korea – Coevolution of International and Domestic Knowledge Linkages[59]

Countries with emerging economies must rely primarily on foreign sources of knowledge as the main vehicle of learning and capability formation. International linkages are needed to pave the way for an effective exploitation of latecomer advantages. Empirical research has shown that, as a developing country progresses in its industrial transformation, its reliance on international technology sourcing and knowledge linkages substantially increases.[60] The Korean innovation system in the electronics industry is emblematic for a heavy reliance on international linkages, combined with the development of complementary domestic linkages.[61]

Early on, as a part of its innovation strategy, the Korean "government encouraged some of the leading chaebol[62] to focus on learning and knowledge accumulation through a variety of links with foreign equipment and component suppliers, technology licensing partners, OEM clients, and minority joint-venture partners."[63] In addition, much of Korea's success lay in its firms' abilities to develop the knowledge and

---

[56]ECFA is a special free-trade agreement between Taiwan and China, which was concluded in September 2010.

[57]For instance, during the November 2011 Asia-Pacific Economic Cooperation meeting in Honolulu, Taiwan supported U.S. proposals to extend the Information Technology Agreement and to establish an Environmental Goods and Services Program.

[58]On SoC design, the government has initiated a National SoC Research Program. On nanotechnology R&D, the government has committed substantial funds, while ITRI and the National Science Council have signed an agreement to conduct joint research with the National Research Council of Canada. And Sha et al. ("ITRI's Role in Developing the Access Network Industry in Taiwan" in H. S. Rowen, M. G. Hancock, and W. F. Miller (eds.), 2008, *Greater China's Quest for Innovation*, Shorenstein APARC, Stanford, CA) describe ITRI's role in the industry-level upgrading of Taiwan's access network industry.

[59]This section heavily relies on Dieter Ernst, "Global Production Networks and the Changing Geography of Innovation Systems: Implications for developing Countries." *East-West Center Working Papers, No. 9, November 2000*. Available at http://scholarspace.manoa.hawaii.edu/bitstream/handle/10125/6074/ECONwp009.pdf?sequence=1. Last accessed August 21, 2012.

[60]For instance, Ernst, Ganiatsos, and Mytelka (eds.), 1998, *Technological Capabilities and Export Success - Lessons from East Asia*, Routledge, London and New York.

[61]D. Ernst, 2000, "Catching-Up and Post-Crisis Industrial Upgrading. Searching for New Sources of Growth in Korea's Electronics Industry," in F. Deyo, R. Doner, and E Hershberg (eds.), *Economic Governance and Flexible Production in East Asia*, Rowman and Littlefield Publishers. Taiwan provides another, albeit very different, approach to the development of network integration services through international linkages.

[62]*Chaebol* refers to South Korean business conglomerates that are global multinationals owning numerous international enterprises.

[63]D. Ernst, 2000, "Catching-Up and Post-Crisis Industrial Upgrading. Searching for New Sources of Growth in Korea's Electronics Industry," in F. Deyo, R. Doner, and E Hershberg (eds.), *Economic Governance and Flexible Production in East Asia*, Rowman and Littlefield Publishers. Taiwan provides another, albeit very different, approach to the development of network integration services through international linkages.

skills necessary to monitor, unpackage, absorb, and upgrade foreign technology. Equally important was a capacity to mobilize the substantial funds for paying technology licensing fees and for importing best-practice production equipment and leading-edge components. Most Korean producers arguably would have hesitated to pursue such high-cost, high-risk strategies had they not been induced to do so by a variety of selective policy interventions by the Korean state. By providing critical externalities such as information, training, maintenance and other support services, and finance, the Korean government has fostered the growth of firms large enough to overcome high entry barriers.

It is this coevolution of international and domestic knowledge linkages that explains Korea's extraordinary success. It has enabled Korean firms to reverse the sequence of technological capability formation. Rather than proceeding from innovation to investment to production, they focused on the ability to operate production facilities according to competitive cost and quality standards.[64]

Through reverse engineering and other forms of copying and imitating foreign technology, as well as integrating into the increasingly complex global production networks of American, Japanese, and some European global flagship corporations, Korean firms were able to avoid the huge cost burdens and risks involved in R&D and in developing international distribution channels.

For Korea, international linkages provided an important initial catalyst for the development of a sufficiently broad portfolio of domestic capabilities that are needed to reap potential benefits of latecomer advantages.

## 3.5 Europe – Integrated EU-wide Innovation Policy Coordination

### 3.5.1 The Seventh Framework Program for Research and Technological Development (FP7)

The 2007–2013 Seventh Framework Program (FP7) for research and technological development is the European Union's main instrument for funding research in Europe.[65] With a total budget of €53.2 billion, the FP7 aims to increase Europe's growth, competitiveness, and employment through initiatives and existing programs that finance grants to research actors all over Europe, usually through cofinancing research, technological development, and demonstration projects. However, access to funding is restricted to organizations based in the European Union. This restrictive approach to international cooperation in science and technology is further emphasized in the European Commission's (EC) 2010 policy document *Innovation Union* to "ensure that leading academics, researchers and innovators reside and work in Europe and to attract a sufficient number of highly skilled third country nationals to stay in Europe."[66]

The 2012 FP7 Work Program is the EC's largest funding package (about €7 billion) under the FP7 so far and will provide funding to EU-based universities, research organizations, and industries, with special attention given to small and medium enterprises. In addition, it is expected to create around 174,000 jobs in the short term and nearly 450,000 jobs and €80 billion growth in gross domestic product (GDP) over 15 years. Since the initiation of the FP7 Program, investment in industrial R&D by the European Union's top 1,000 companies has grown by ~10 percent. Between 2010 and 2011, industrial R&D in the European Union grew by almost 6 percent, compared with higher growth reported for the United States (~10 percent), Taiwan (~18 percent), Korea (~21 percent), Hong Kong (~29 percent), and China (~30 percent), and lower reported growth in Japan (-10 percent). These industrial R&D investments can also be broken down by industry classification—of specific interest to this study are the R&D contributions from the telecommunications, software, computer science, semiconductors, and electronics industry. These sectors make up ~23 percent of the European Union's[67] and of Japan's industrial R&D investments, compared with ~35 percent for Hong Kong, ~39 percent for China, ~41 percent for the United States, ~41 percent for India,[68] ~70 percent for Korea, ~77 percent for Singapore, and a staggering ~94 percent for Taiwan.[69,70]

In R&D intensity, however, the European Union continues to lag behind Japan and the United States. At

---

[64]Ibid.

[65]See http://ec.europa.eu/research/fp7/index_en.cfm. Last accessed on January 7, 2012.

[66]EC, 2010, *Europe 2020 Flagship Initiative Innovation Union. Communication from the Commission to the European Parliament, the Council, the European Economic and Social Committee and the Committee of the Regions.* Available at http://ec.europa.eu/research/innovation-union/pdf/innovation-union-communication_en.pdf, page 27. Last accessed on January 7, 2012.

[67]Among the European Union member states, Germany, France, and Finland made the largest R&D investments in the telecommunications, software, computer science, semiconductors, and electronics industry sectors (16.1 percent, 22.4 percent, and 82.8 percent, respectively).

[68]India's R&D investment sectors consist primarily software and computer services.

[69]See http://iri.jrc.es/research/scoreboard_2011.htm. Last accessed on January 7, 2012.

[70]See http://iri.jrc.ec.europa.eu/research/scoreboard_2008.htm. Last accessed on January 7, 2012.

1.6 percent, the European Union's 2010 share of R&D expenditure in GDP trails both Japan and the United States by a considerable margin, with 3.3 percent and 2.7 percent shares, respectively.[71] Among the member states, Germany dominates—at 2.5 percent, its share of R&D expenditures in GDP is much larger than the European Union share. More importantly, however, Germany was deemed to have the highest propensity to innovate.[72,73] Hence, it is important to emphasize that national innovation policies differ quite substantially across Europe, both in their overall strategic vision, and in their effectiveness.

*3.5.2. Toward an Integrated EU-wide Innovation Strategy*

Germany's move toward an integrated innovation strategy[74] is emblematic for a growing trend within the European Union to adopt a much more centralized approach to innovation. In 2000 the European Union established the European Research Area (ERA) to promote a "single innovation market." One of its main objectives was to optimize and open European, national, and regional research programs to support the best research throughout Europe and to coordinate these programs to address major challenges together.[75]

In December 2008 the Competitiveness Council adopted a 2020 ERA vision, which seeks to increase the Europe-wide mobility of innovation capabilities by promoting the free circulation of researchers, knowledge, and technology. In 2010 the EC developed an integrated innovation strategy entitled "Innovation Union" to tackle three main challenges for EU innovation policy: (1) underinvestment in knowledge foundation (e.g., the United States and Japan are out-investing Europe and China is rapidly catching up); (2) unsatisfactory framework conditions, ranging from poor access to finance, high costs of intellectual property rights (IPR) to slow standardization and ineffective use of public procurement; and (3) too much fragmentation and costly duplication.[76]

*3.5.3 The European Union's Key Enabling Technologies (KET) program*

An interesting attempt to operationalize Europe's integrated innovation strategy is the European Union's Key Enabling Technologies (KET) Program.[77] The EC's six KETs—nanotechnology, micro- and nanoelectronics, advanced materials, photonics, industrial biotechnology, and advanced manufacturing systems[78]—were selected based on their economic potential, their value-adding and enabling role, and their technology and capital intensity with R&D and initial investment costs. KETs are defined as "knowledge and capital-intensive technologies associated with high research and development (R&D) intensity, rapid and integrated innovation cycles, high capital expenditure and highly-skilled employment."[79] KETs are also embedded in advanced products and they underpin innovation chains.

Advanced computing products, such as multicore processors and parallel software developments, are examples of technologies that are consistent with the KET definition. Like other KETs, advanced computing technologies provide potential first-mover advantages, and enable the owner of relevant intellectual property rights to create new lead markets as new technologies replace old technologies with few or no other players. One of the key goals of the European Union's KET Program is to reduce the deeply ingrained barriers to industrial innovation. In other words: Why are breakthrough ideas, inventions, and discoveries (that

---

[71]Battelle, 2010, "Global R&D Funding Forecast" in *R&D Magazine*, December 2009. Available at http://www.rdmag.com/uploadedFiles/RD/Featured_Articles/2009/12/GFF2010_ads_small.pdf. Last accessed on August 11, 2012.

[72]An innovation, here, is defined as a new or significantly improved product (good or service) introduced to the market or the introduction within an enterprise of a new or significantly improved process.

[73]Eurostat, *2010 Yearbook*, p. 606. Available at http://epp.eurostat.ec.europa.eu/cache/ITY_OFFPUB/KS-CD-10-220/EN/KS-CD-10-220-EN.PDF. Last accessed on August 15, 2012.

[74]Federal Ministry of Education and Research (BMBF) (2011) High-Tech Strategy. Available at http://www.hightech-strategie.de/en/350.php. Last accessed on January 12, 2012.

[75]See http://ec.europa.eu/research/era/index_en.htm. Last accessed on January 12, 2012.

[76]EC, 2010, *Europe 2020 Flagship Initiative Innovation Union. Communication from the Commission to the European Parliament, the Council, the European Economic and Social Committee and the Committee of the Regions,* Commission Communication (COM(2010)546). Available at http://ec.europa.eu/research/innovation-union/pdf/innovation-union-communication_en.pdf, page 27. Last accessed on January 7, 2012.

[77]EC, 2011, *High-Level Expert Group on Key Enabling Technologies. Final Report*, June. Available at http://ec.europa.eu/enterprise/sectors/ict/files/kets/hlg_report_final_en.pdf. Last accessed on August 15, 2012.

[78]EC, 2009, *Preparing for our future: Developing a common strategy for key enabling technologies in the EU,* Commission Communication (COM(2009)512). Available at http://ec.europa.eu/enterprise/sectors/ict/files/communication_key_enabling_technologies_en.pdf. Last accessed on August 15, 2012.

[79]EC, 2010, *Current situation of key enabling technologies in Europe*, Commission Staff Working Document (SEC(2009)1257). Available at http://ec.europa.eu/enterprise/sectors/ict/files/staff_working_document_sec512_key_enabling_technologies_en.pdf. Last accessed on August 15, 2012.

were developed with public R&D funds) not transformed into commercially successful innovations within reasonably short time frames?

*3.5.4 Policy Options*

To overcome the above deeply entrenched innovation barriers, the European Union's KET Program proposes a broad range of coordinated support policies that cover the following stages of the "innovation chain," from the transformation of fundamental research into globally competitive technologies, through product development to make innovative and cost-effective product development and prototyping, to globally competitive manufacturing.

Specifically, the EC KET Program identifies the following five priority areas for Europe's evolving EU-wide innovation strategy: (1) sustain a critical mass in knowledge and funding through effective use of economies of scale and scope; (2) increase market focus of R&D projects; (3) invest in large-scale demonstrators and pilot test facilities; (4) provide post-R&D commercialization support; and (5) practice trade diplomacy, that is, reduce unfair subsidies and protect domestic companies from unfair trade practices.[80] This last policy priority is of particular concern from a U.S. perspective. In fact, the European Union's KET Program culminates in a fairly "techno-nationalist" notion of IPR protection and states that "the EU should clearly promote an 'in Europe first' IP policy" and that proposals require clear IP plans for "first exploitation of IP" and rules that "favour the EU exploitation of the results of projects."[81]

*3.5.5 Summary*

The European Union has experienced a fundamental change in its innovation policy from government-centered national strategies to attempts to combine market-led innovation and public policy coordination across Europe. While government initiatives, such as the KET Program, attempted to bridge the perennial gaps that stymie Europe's industrial innovation ecosystem, significant challenges remain. To a large degree, however, this transformation is still a work in progress, as European IT innovation and commercialization continue to lag.

In addition, there are signs that Europe's fiscal crisis and increasingly severe austerity policies might slow down Europe's move towards greater openness and internationalization of its innovation system.

---

[80]EC, 2011 KET, p. 33.
[81]EC, 2011 KET, p. 37.

## 3.6 Conclusions and Policy Implications

The diversity of economic and IT innovation policies across the United States, China, Taiwan, and Europe reflect their differing cultures and history, economic status and technical capabilities. The U.S. approach rests on government support for basic academic research and a vibrant capital market and private enterprise ecosystem for product innovation. The other countries and regions blend elements of private enterprise and central planning. Each is unique and not directly transferrable to another region. Nevertheless, there are some general principles that can be gleaned from this survey of policies, coupled with technical insights regarding semiconductor device fabrication, chip architecture, and software.

Some of the largest computing companies in the United States have internal multidimensional technological capabilities in chip design, process development, wafer manufacturing, and software and have demonstrated success tapping into foreign talent pools and markets. However, IT talent, capabilities, and facilities are increasingly distributed globally. Although research prowess is correlated with industry success, information flows globally via many sources. The lesson of basic research, both in industry and academia, has been that the discoverers are not always those who convert the ideas into economically successful products. Oftentimes, the likelihood that an idea can be successfully commercialized and implemented depends on a nation's or region's innovation policies and entrepreneurial climate.

Second, the cost of semiconductor fabrication facilities is rising exponentially, placing their construction beyond the economic reach of small- and mid-sized companies. Only the largest multinational companies and nation-states can fund their construction and operation. This suggests that the United States must be mindful of its global dependence on fabrication supply chains and that it develop realistic models that balance the need for the latest process technology versus multiaxis innovation and that combine reliability and resilience, programmability, and functionality. Although financial investment in fabrication facilities by a small number of U.S. companies, primarily by Intel, provides some domestic sourcing, most of the chips contained in devices sold in the United States are fabricated offshore. IBM does produce some chips in the United States, both for U.S. defense needs and its own products, but the volume is relatively small.

Third, there is no assurance that historical U.S. dominance in computing will transfer to new and emerging domains. The need for architectural and software innovation to deliver new features and greater performance via parallelism creates opportunities for new ecosystems to emerge and evolve. With licensable components and global access to fabrication facilities, it is possible for this innovation to occur almost anywhere. In addition to performance as measured by computing speed (clock speed, bandwidth, interconnect, and so on), it may be that other measures—such as reliability and resilience, programmability, security, and efficiency—become equally, or potentially more, important. For example, efforts to improve programmability and efficiency of base processors might yield significant improvements in software quality, software development times, and (ultimately) application performance.

Fourth, global policy makers see information technology in general and consumer computing in particular as major economic forces to be harnessed for local and regional benefit. They are investing in the future, hoping to position their region for success. Which of the myriad approaches being pursued will be most successful is difficult to predict.

Today is an inflection point, when the virtuous cycle of faster sequential processors has broken down and when new devices and services are emerging to reshape the computing landscape. An intense global competition for IT hegemony is under way. No company, country, or region will reap all of the economic benefits, as the global value chain is too intertwined for that. However, there will be economic winners and losers, just as there always are whenever technology shifts occur. U.S. policy makers would be wise to think carefully and deeply about the shifts under way and their implications for economic competitiveness and national security.

# 4

# Implications of Changes in the Global Advanced Computing Landscape for U.S. National Security

The viability, efficiency, and, ultimately, success of the global economic ecosystem depend in part on the flow of goods and services throughout the global value chain, from design through fabrication to consumption. In computing and information technology, the locus of innovation, influence, and early access can and has shifted throughout the history of the modern information technology (IT) era. These shifts can have significant implications for U.S. competitiveness and national security.

In the committee's view, the United States currently enjoys a technological advantage in advanced computing hardware and software capabilities but that technological gap is narrowing, not only due to the technical challenges described in Chapter 1, but also because other global economic competitors (e.g., China) are making a concerted effort to develop their own indigenous computing design and manufacturing capabilities. Moreover, the design and fabrication of such technologies are increasingly globally distributed. Market success, of course, is only partially correlated with technological preeminence, as the ecosystem of producers and consumers and market size, together with network effects, are also key determinants.

In the committee's view, national security concerns for the United States related to anticipated long-term developments in advanced computing come not just from potential threats to U.S. technological superiority, but also from changes to the nature and structure of technical innovation and to the marketplace for computing and information technology. Intensifying competition will affect the global supply chain and reshape the numbers and types of commercial players that survive in a rapidly evolving marketplace. The diminishing performance returns from traditional silicon advances that have helped existing software systems run ever faster (described in Chapter 1) and the rise of the post-personal computer (PC) ecosystem of smart devices, coupled with cloud-computing capabilities, further complicate the landscape. This chapter discusses several emerging changes in the global advanced computing landscape that have implications for U.S. national security, including parallelism in hardware and software (Section 4.1), the integrity and reliability of the global supply chain (Section 4.2), the decline of custom production (Section 4.3), convergence of civilian and defense technological capabilities (Section 4.4), the rise of a new post-PC paradigm driven by mass information and communications technology (ICT) consumerization (Section 4.5), new market-driven innovation centers (Section 4.6), the future educational and research landscape on advanced computing (Section 4.7), cybersecurity and software (Section 4.8), and possible defense ICT outcomes (Section 4.9).

## 4.1 Parallelism in Hardware and Software

In U.S. defense and national security, one element of the U.S. advantage in defense ICT has accrued from rapid increases in application performance, which in turn has depended on rapid increases in single-processor performance. As the latter ends, continued application performance increases will likely only be possible if there is a shift to the development of applications that can take advantage of parallel hardware. The inability of defense software to make this transition faster than competitors in the global market or our potential adversaries has significant implications for U.S. competitiveness and national security.

The slowdown in performance increases for single-core processors is a matter of physical (e.g., power density and dissipation and quantum barriers) and technological (e.g., gate length, lithography, power dissipation, wire scaling, and materials) limits. Thus, even if resources were plentiful, it would not be possible to simply buy or make appropriately targeted investments that would result in continuing exponential speed-ups for single-core processors. An additional matter of physics is the power constraints that are driving the industry from homogeneous multicore chips to heterogeneous parallelism, for example, using graphics processing units, accelerators, and reconfigurable fabrics. Once again, national security processes and deployments will need to adapt to use heterogeneous parallelism to maintain advantage.

Exponentially increasing processor speed has traditionally served as a proxy for higher-performing, more capable, and more innovative systems. Absent this traditional metric of continually increasing performance (whether from sequential or parallel systems), focusing on other metrics will likely come to the fore. In many cases, design and innovation efforts will focus on combinations of improvements in diverse dimensions, such as cost, energy, weight, robustness, and security. Traditional performance improvements would help to achieve these, but if such improvements are not forthcoming, other means of achieving these improvements will be needed.

Developing, verifying, and deploying software to complement advanced hardware is fraught with challenges. Moves to homogeneous and then heterogeneous parallelism will amplify these challenges.[1] These challenges are especially prevalent in defense and national security systems. Moreover, defense is notable for its relatively slow adoption of innovative hardware and software that now emerge from the commercial rather than the military sector. New and faster-moving threats with fewer legacy concerns may make this status quo untenable.[2,3]

U.S. national security has long relied on an information technology advantage. Given the dramatic shift to multicore chips and explicit parallelism, defense ICT will need to transition to tools, techniques, and processes that can meet defense needs through effective use of new parallel software models and emerging hardware approaches. Such a transition will be difficult, and even if this transition is made successfully—a challenge not just for defense, but for even the most advanced commercial interests as well—the growth rate of computing performance is expected to continue to slow, making it easier for the rest of the world, including adversaries, to catch up.

## 4.2 Integrity and Reliability of the Global Supply Chain

Maintaining the integrity of the global supply chain is a serious challenge. The supply chain for integrated circuits (computer chips) is of particular interest given that they are key components of all computing systems. Some fabrication facilities are still present in the United States. For instance, Intel is the primary operator of large-scale, state-of-the-art semiconductor fabrication facilities in the United States, though it also has such facilities outside the United States. IBM and other companies operate facilities in the United States that target more specialized markets and national security needs.[4] However, the United States is increasingly dependent on foreign sources of microchip production and on device assembly and testing capabilities that are concentrated in a handful of countries.

Developing secure sources of production is also challenging. A global supply chain increases the likelihood that compromised and counterfeit products can be introduced in mission-critical infrastructure.[5]

---

[1] The 2009 National Research Council (NRC) report *Critical Code: Software Producibility for Defense* assesses the growing importance of software for national security and examines how the U.S. DOD can most effectively meet its future software needs.

[2] The 2009 NRC report, *Achieving Effective Acquisition of Information Technology in the Department of Defense* calls for the DOD to acquire information technology systems using a fundamentally different acquisition process based on iterative, incremental development practices.

[3] This is reminiscent of the 20th century U.S. automobile sector. The U.S. auto industry moved from being the best in the world to being high cost and slow to adopt new processes and technologies. This decline was masked for years by the lack of credible competition. The arrival of Japanese and other foreign automakers changed the competitive landscape two ways. First, the Japanese focus on manufacturing efficiency exposed U.S. companies' process problems and eroded near-term profits. Second, sustained long-term Japanese investments (e.g., on energy efficiency) con-

trasted with U.S. companies' more near-term focus. Too much of a short-term focus cuts into long-term success. Moreover, once the former made less money available, addressing the latter became more difficult.

[4] See http://www.nsa.gov/business/programs/tapo.shtml. Last accessed on July 2, 2012.

[5] Indeed, an immediate challenge for U.S. access to the global semiconductor value chain is that some U.S. defense contractors have been deceived into using counterfeit electronics parts. At a November 2011 hearing, the Senate Armed Services Committee noted that such fake parts could have disastrous consequences for the performance of U.S. defense equipment such as helicopter night-vision systems and aircraft video display units. See U.S. Senate Committee on Armed Services, *Hearing to receive testimony on the Committee's investigation into counterfeit electronic parts in the Department of Defense supply chain,* November 8, 2011. See also DOD's TRUST in Integrated Circuits Program (available at (http://www.darpa.mil/Our_Work/MTO/Programs/

Developing ways to ascertain and monitor the provenance of semiconductor products will become ever more important. Related issues of hardware and software verification and validation will continue to be critical issues, particularly as the complexity of systems continues to rise.

In addition to challenges related to integrity and security, the global interdependence of design, component fabrication, and assembly means that risks of disruption due to natural disasters, political conflict, or constrained access to raw materials become greater. A single event, such as the March 2011 earthquake-tsunami in Japan or the more recent floods in Thailand, can disrupt global product deliveries for months.[6] Similarly, restrictions on shipments of rare earths, key elements of chip fabrication, can stall production lines. The globalization of science and technology (S&T) and of the computing marketplace in combination with specialization (only a few suppliers of a particular component) and just-in-time inventory practices all add to the risk as well. More generally, a disaster or a well-targeted action from an adversary could constrain or interrupt global supplies, potentially placing the United States in a defensive position due to competing demands between U.S. defense needs, commercial production requirements, and the producing region's own needs.

## 4.3 Decline of Custom Production

A decrease in the number of specialized companies able to make custom products for defense needs is also relevant to national security. Although commercial off-the-shelf (COTS) products are widely used in defense materials, there are specialized components and products that are not commodity products.

This reduction is driven in part by the exponentially rising cost of state-of-the-art fabrication facilities, which places a premium on volume production. In turn, this limits the economic incentive for any company to respond to the defense needs for specialized devices—for example, the capacity to design and fabricate radiation-resistant integrated circuits (ICs). Further, the concentration of design and production to a small group of dominant market players that make commoditized products may significantly increase costs.

## 4.4 Convergence of Civilian and Defense Technological Capabilities

The convergence of civilian and defense technologies is accelerating, driven by rapid and cost-effective technological progress in a highly competitive commercial marketplace, especially as compared with the often lengthy and rigid procurement processes in the defense sector. Convergence is most evident in electronics, where a growing proportion of U.S. defense needs are being met by COTS technologies. At the same time, the U.S. defense establishment's ability to influence the development of the global semiconductor industry, similarly to what happened with supercomputers (which create the chip components of COTS products) through sheer volume has been reduced. For instance, the U.S. military accounted for a large proportion of sales from the global semiconductor industry in that industry's formative years, but that proportion had fallen to just 1 percent of global microcircuit sales by the late 2000s.[7]

The convergence between civilian and defense hardware capabilities and ease of access to openly available technological products that may be just as good or even more advanced than equivalent defense technologies has implications for U.S. defense.[8] In particular, such convergence allows greater opportunity for adversaries to narrow the technological gap with the United States. In such an environment, time to integration and time to deployment will be the primary factors that determine technical superiority, rather than who is the first to develop a particular technology.

This suggests that deeper awareness of the differing processes and timescales for hardware and software development must be part of the design and procurement process. Semiconductor design and fabrication, as well as subsequent integration of fabricated chips, have a substantial lead time. Although it is possible to develop portions of new software systems with simulators and emulators, integration and complete testing is dependent on hardware availability. Thus, the overall time to deployment of new hardware and software systems will be especially critical when the software requirements for

---

Trusted_Integrated_Circuits_%28TRUST%29.aspx, last accessed on February 7, 2012) that seeks to "provide trust in the absence of a 'trusted foundry'."

[6]As an example, the shortages of disk drives and flash memory resulting from the Japan and Thailand natural disasters affected many devices and vendors. See http://www.isuppli.com/Home-and-Consumer-Electronics/News/Pages/IHS-iSuppli-News-Flash-Thailand-Flood-Spurs-Nearly-4-Million-Unit-Shortfall-in-PC-Shipments-in-Q1-2012.aspx Last accessed on February 7, 2012.

[7]*Annual Industrial Capabilities Report to Congress, 2008* (Washington D.C.: Office of Under Secretary of Defense Acquisition, Technology and Logistics Industrial Policy, February 2008).

[8]An ongoing NRC study, *Ethical and Societal Implications of Advances in Militarily Significant Technologies that are Rapidly Changing and Increasingly Globally Accessible* is exploring these issues.

defense missions must be developed based on predicted, rather than current hardware. This further emphasizes the importance of hardware-software co-design and rapid testing and of drawing lessons from consumer device deployment.

The expected length of the life cycle for consumer devices continues to decrease; for instance, the replacement time for smartphones is now less than a year. Comparatively slow and cumbersome Department of Defense (DOD) procurement and deployment cycles mean that units may lack access to current-generation technology. Defense organizations must balance rapid adoption for commodity technologies against more measured and careful integration and deployment of devices and technologies that are unique to defense needs. Risks increase when applying the same process and evaluation to both without distinguishing the risks and benefits.[9] At the same time, proven technology— even if it is not the most current—may provide better results with cost-effective performance. Managing these tensions suggests that requirements and designs should be based not just on current technology but on projections of technology available two or even three generations ahead.

## 4.5 Rise of a New Post-PC Paradigm Driven by Mass ICT Consumerization

One area in which COTS has become the principal technological driver is in the ongoing consumerization of ICT and the emergence of what might be called a post-PC technological paradigm. Smartphones, tablets, cloud-computing capabilities, and other related commercial technologies are the hallmarks for this new era. Industry projections[10] suggest there could be as many as 50 billion devices connected to the Internet within a decade. Global sales of mobile phones now exceed those of PCs, and the Chinese phone market alone exceeds that of the United States or Europe.[11] For much of the world's population, a phone is the primary computing device.

More generally, low-power designs, based on licensable components and created by semiconductor design firms without fabrication capabilities, along with the rise of system-on-a-chip (SoC) ecosystems are increasingly enabling new companies and enterprises to offer devices that compete with the traditional x86-oriented PC ecosystem.

In both the x86 and ARM SoC ecosystems, some elements of each SoC are likely to be common (for example, general-purpose cores); others will be tailored to specific applications (for example, cryptography blocks, media encoders and decoders, digital signal processors, or network interfaces) and drawn from an array of internationally available and licensable silicon design blocks. This mix-and-match model, now prevalent in the mobile device space, challenges the traditional software development and maintenance model, where legacy software could execute unchanged (often without recompilation) as described in Chapter 1. A DOD shift to application-tailored classes of chips will require software refactoring and optimization for each new class of chips, each with different functionality, adding complexity to the software design and maintenance life cycle. Unless the software design process and toolset for distinctive defense software is adapted to this shift, the useful lifetime of the chips will be determined by software availability, not hardware.

In addition to the rise of a new and complementary COTS ecosystem, the consumerization of ICT has profound implications for how organizations manage their own ICT. The proliferation and popularity of new device functionality challenges traditional approaches to organizational technology uptake. Consumers drive adoption of technology in large organizations by forcing central ICT organizations to respond to consumer acquisition outside the organization. This socially activated disruption changes the planning and deployment of software and services. The DOD is not immune to this effect. As the perceived and actual differences between commodity technology availability and centrally mandated deployments rises, individuals and groups may circumvent best policies and practices in system security and information flow in order to access improvements in functionality.[12] In addition, the proliferation of mobile devices with personally identifiable data and institutional data brings information leakage risks due to the possibilities of device loss and theft.

---

[9] A 2009 NRC report, *Achieving Effective Acquisition of Information Technology in the Department of Defense* calls for the DOD to acquire information technology systems using a fundamentally different acquisition process based on iterative, incremental development practices.

[10] CISCO White Paper, 2011, *The Internet of Things: How the Next Evolution of the Internet is Changing Everything*. Available at http://www.cisco.com/web/about/ac79/docs/innov/IoT_IBSG_0411FINAL.pdf. Last accessed on February 7, 2012.

[11] See http://www.strategyanalytics.com/default.aspx?mod=reportabstractviewer&a0=6871. Last accessed on February 7, 2012.

[12] A recent NRC report, *Toward Better Usability, Security, and Privacy of Information Technology,* examines some of the competing motivations for users of technology and identifies research opportunities and ways to embed usability considerations in design and development related to security and privacy, and vice versa.

Similarly, the rise of big data and rich data analytics, made possible by the proliferation of these inexpensive networked devices and by massive cloud data centers, are challenging traditional notions of computing. The balance of value is shifting from isolated devices and software to capabilities embodied by an integrated system of devices, data, and services. A data-rich, consumer-driven world where data is ubiquitous and often accessible has profound implications for U.S. DOD notions of information superiority, privacy, and security.

The increasing diversity and independence of global supply chains for new generations of COTS devices will challenge existing approaches to system security. Demand for software verification of diverse components with multiparty provenance will increasingly be the norm, but thus far verification of even existing systems remains a challenging research problem. When coupled with device heterogeneity and specialization for performance, verifying functionality and the absence of implicit or explicit security backdoors will require new organizational and software security approaches.

## 4.6 New Market-Driven Innovation Centers

The emergence of foreign markets that are larger, are potentially more lucrative, and have better long-term growth potential than in the United States and other developed countries also has significant implications for the ability of the United States to shape technological directions. A shift in the global commercial center of gravity (either as the result of a new development or of decreased public or private research investments) may lead to a shift in the global research and development (R&D) center of gravity. For example, this could occur if international firms are required to locate in these markets to remain competitive, to meet the requirements of government regulations in the target markets, and to better understand those markets. The availability of trained and talented researchers and developers, particularly in parallel computing, will also affect these placements, as today's devices are dependent on parallel applications and system software to meet performance and functional targets.

## 4.7 The Future Educational and Research Landscape in Advanced Computing

In the committee's view, the United States became the leader in advanced computing because of its significant and sustained investment in long-term basic research, especially its combination of risky, big bets, some of which had significant financial returns, and curiosity-driven, smaller-scale research.[13] A diverse portfolio of research was supported by multiple agencies in the Networking and Information Technology Research and Development (NITRD) Program over many years. This era of diverse funding has undergone changes in recent years. There is now an increasing monoculture of research funding for computing research, centered on the funding model of the National Science Foundation (NSF). Because NSF emphasizes single-investigator and small-group research, it has not historically supported long-term, large-scale infrastructure for either chip and system fabrication or compiler and tool infrastructure.

In contrast, China, for example, has been increasing its R&D investment in advanced computing over the past decade and appears willing to invest in research aimed at both incremental and higher-end computer innovation. At the same time, China is investing heavily in the training of advanced scientists and engineers at the undergraduate and postgraduate level both at home and abroad. If these trends continue, the still-wide gulf in the educational and R&D capabilities between China and the United States will narrow.

## 4.8 Cybersecurity and Software

The growing R&D competitiveness of other countries has potentially far-reaching ramifications for the United States in cybersecurity. The DOD and the U.S. government cybersecurity strategy depends upon the U.S. commercial information technology sector remaining as the world leader.[14] Software development is an increasingly central driver of computing innovation, whether it is parallel tools and applications for new devices or advanced software services and data analytics running atop massive, highly parallel cloud data centers.

The interconnected nature of globally designed and manufactured consumer devices contributes to increased risk of data and software security breaches and makes clean separation of functions—a traditional tenet of good security—ever more difficult. The globalization of this software development, as well as state-sponsored cyber-espionage, raises important software and cybersecurity questions.[15] Cybersecurity may well become a pivotal long-term area of competition between the United States

---

[13] NRC, 1999, *Funding a Revolution: Government Support for Computing Research*, Washington, D.C.: The National Academies Press (available online at http://books.nap.edu/catalog.php?record_id=6323).

[14] Office of the Deputy Assistant Secretary of Defense for Industrial Policy, 2010, *DoD Cyberstrategy: Leveraging the Industrial Base*, December.

[15] The NRC has a deep portfolio of work on cybersecurity: http://www.nas.edu/cybersecurity.

and foreign competitors with fast-growing software industries, most notably India and China.[16]

### 4.9 Possible Defense IT Outcomes

The slowdown in the growth of single-processor computing performance described in Chapter 1 brought an end to the virtuous cycle of ever-faster sequential processors coupled with increasingly feature-rich software built atop a sequential model. Explicit parallelism in both hardware and software is now required to realize greater performance and desired functionality. The consequences of this shift are deep and profound for computing and for the sectors of the economy that depend on and assume, implicitly or explicitly, ever-increasing performance. From a technology standpoint, this has lead to heterogeneous multicore chips and a shift to new innovation axes that include but are not limited to chip performance. In turn, these technical shifts are reshaping the computing industry, with global consequences.

Today, global equilibration, access to standard hardware Internet protocol (IP) blocks, and open foundries have lowered the barrier to entry for international competitors, particularly in Asia. As a result, it is possible that the locus of innovation may shift further from the United States. Technology limitations are forcing a new ecosystem of mix-and-match IP blocks and heterogeneous multicore SoCs on all computer systems. This trend and the proliferation of device types present daunting challenges, especially given the historical hegemony of the United States in mainstream computing. Barring concerted action involving major technology breakthroughs and a major shift in U.S. industrial competitive policy, this accelerating innovation shift may open the door to a latecomer innovation advantage (discussed in Chapter 3).

The challenges and the opportunities for the United States are in capitalizing on its historical strengths in systems design, engineering, and integration. Defense systems and their information technology components are often large and complex, with interconnected and often redundant components. Advanced computing is a critical element of such systems, but only one element. If the United States focuses on nimble and rapid system integration, with designs that emphasize reliability and verification, it can continue to build effective defense systems.

Otherwise, the DOD could find itself with deployed computing technology that is no better than, or even inferior, to its adversaries.[17] Such technical parity (or even inferiority) could occur due to either a loss of U.S. technological capabilities or the inability to deploy the appropriate new technologies sufficiently rapidly to maintain a competitive advantage.

---

[16]N. Gregory, S. Nollen, and S. Tenev, 2009, *New Industries from New Places: The Emergence of the Software and Hardware Industries in China and India*, Stanford University Press and World Bank, Washington, D.C.

[17]Further, computing technologies could also potentially be manufactured by adversaries.

# Appendixes

# A

# Committee Member Biographies

**DANIEL REED** is Vice President for Research and Economic Development at the University of Iowa, where he holds the University Computational Science and Bioinformatics Chair, with joint appointments in Computer Science, Electrical and Computing Engineering and Medicine. Previously, he was Microsoft's corporate vice president for technology policy. Prior to that, he was Chancellor's Eminent Professor at the University of North Carolina (UNC) at Chapel Hill, as well as the director of the Renaissance Computing Institute (RENCI) and the Chancellor's Senior Advisor for Strategy and Innovation for UNC Chapel Hill.

Dr. Reed has served as a member of the U.S. President's Council of Advisors on Science and Technology (PCAST) and as a member of the President's Information Technology Advisory Committee (PITAC). As chair of PITAC's computational science subcommittee, he was lead author of the report *Computational Science: Ensuring America's Competitiveness*. On PCAST, he cochaired the Subcommittee on Networking and Information Technology (with George Scalise of the Semiconductor Industry Association) and coauthored a report on the National Coordination Office's Networking and Information Technology Research and Development (NITRD) Program called *Leadership Under Challenge: Information Technology R&D in a Competitive World*. In June 2009 he completed two terms of service as chair of the board of directors of the Computing Research Association, which represents the research interests of Ph.D.-granting university departments, industrial research groups and national laboratories.

He was previously head of the Department of Computer Science at the University of Illinois at Urbana-Champaign (UIUC), where the held the Edward William and Jane Marr Gutgsell Professorship. He has also been director of the National Center for Supercomputing Applications (NCSA) at UIUC, where he also led the National Computational Science Alliance, a 50-institution partnership devoted to creating the next generation of computational science tools. He was also one of the principal investigators and chief architect for the National Science Foundation (NSF) TeraGrid. He received his B.S. from Missouri University of Science and Technology and his M.S. and Ph.D. in computer science in 1983 from Purdue University. He is a fellow of the Association for Computing Machinery (ACM), the Institute of Electrical and Electronics Engineers (IEEE), and the American Association for the Advancement of Science (AAAS).

**CONG CAO** is one of the leading scholars in the study of science, technology, and innovation in China. He is currently an associate professor and reader at the School of Contemporary Chinese Studies, University of Nottingham. Having studied in both China and the United States and in both natural and social science, he received his Ph.D. in sociology from Columbia University in 1997 and has worked at the University of Oregon, the National University of Singapore, and the State University of New York.

Dr. Cao is interested in the social studies of science and technology with a focus on China. He is the author of *China's Scientific Elite* (London and New York: RoutledgeCurzon, 2004), a study of the Chinese scientists holding honorific membership in the Chinese Academy of Sciences, and *China's Emerging Technological Edge: Assessing the Role of High-End Talent* (with Denis Fred Simon, Cambridge and New York: Cambridge University Press, 2009). His journal publications have appeared in *Science*, *China Quarterly*, *Asian Survey*, and *Minerva*, among others.

**TAI MING CHEUNG** is an associate research scientist at the University of California Institute on Global Conflict and Cooperation (IGCC) located at the University of California, San Diego (UCSD), in La Jolla. He directs the Minerva program on Chinese security and technology, a multiyear academic research and training project funded by the U.S. Defense Department to explore China's technological potential. His responsibilities include managing the institute's Track II Program: the Northeast Asia Cooperation Dialogue, which brings together senior foreign ministry and defense officials as well as academics from the United States, China, Japan, South Korea, North Korea, and Russia for informed discussions on regional security issues.

Dr. Cheung is also an associate adjunct professor at UCSD's Graduate School of International Relations and Pacific Studies (IR/PS), where he teaches courses on Asian security, Chinese security and technology, and Chinese politics.

Dr. Cheung is a long-time analyst of Chinese and East Asian defense and national security affairs, especially defense economic, industrial, and science and technological issues. His latest book, *Fortifying China: The Struggle to Build a Modern Defense Economy*, was published by Cornell University Press in 2009. The book examines the economic, commercial, and technological foundations of China's long-term defense modernization that examines the development of the defense industrial complex, the role and prospects for civilian-military integration, and the military dimensions of science and technology policies. He was based in Northeast Asia (Hong Kong, China, and Japan) from the mid-1980s to 2002 covering political, economic, and strategic developments in Greater China and East Asia as a journalist for the *Far Eastern Economic Review* from 1988–1993 and subsequently as a political and business risk consultant for a number of companies, including PricewaterhouseCoopers. Dr. Cheung received his Ph.D. in war studies from King's College London in 2007.

**JOHN CRAWFORD** is an Intel fellow, Digital Enterprise Group, and sets the architectural direction for emerging power and reliability technologies for future Intel processor server platforms. When Crawford joined Intel as a new college graduate in 1977, he worked as a software engineer developing software tools for Intel's 8086 processor including the code-generation phase of Intel's Pascal compiler for the 8086. In 1982 he became the chief architect for the Intel386 microprocessor. He was responsible for defining the company's 32-bit architectural extensions to the already successful 8086/186/286 16-bit product line. In this capacity, he set the architectural direction and later participated in the design of the processor by leading the microprogram development and test program generation. Mr. Crawford made similar contributions as chief architect of the Intel486 processor. He comanaged the design of the Pentium processor from inception through a successful product launch in 1993. Mr. Crawford headed the joint architecture research with Hewlett-Packard that developed the Itanium family architecture, Intel's 64-bit Enterprise product line. He has been involved with the Itanium family of products since its inception in 1994. In 1995, he received the ACM/IEEE Eckert-Mauchly Award for contributions to computer and digital systems architecture, and in June 1997 he received the IEEE Ernst Weber Engineering Leadership Recognition. Mr Crawford was elected to the National Academy of Engineering in 2002.

Mr. Crawford received a bachelor's degree in computer science from Brown University in 1975, and a master's degree in computer science from the University of North Carolina at Chapel Hill in 1977. He holds 23 patents.

**DIETER ERNST** (senior East-West Center fellow at the full professional level) is an authority on global production networks and research and development (R&D) internationalization in high-tech industries and on industrial and innovation policies in China, the United States and emerging economies, with a focus on standards and intellectual property rights. Earlier positions include senior advisor to the Organization for Economic Cooperation and Development (OECD), Paris; research director of the Berkeley Roundtable on the International Economy (BRIE) at the University of California, Berkeley; and professor of international business at the Copenhagen Business School.

Dr. Ernst has cochaired an advisory committee of the U.S. Social Science Research Council to develop a program on innovation, business institutions and governance in Asia. He has served as scientific advisor to governments, private companies, and international institutions, such as World Bank, the Organization of Economic Cooperation and Development, the UN Conference on Trade and Development and the UN Industrial Development Organization. In the United Sates, Dr. Ernst has served as advisor to the National Science Foundation, Social Science Research Council, U.S.-China Economic and Security Review Commission, Council on Foreign Relations, the National Bureau for

Asian Research, U.S. Department of Commerce, the Deloitte Center for the Edge, and the Frontier Strategy Group.

Relevant publications include *Indigenous Innovation and Globalization: The Challenge for China's Standardization Strategy* (2011) [now published in Chinese]; *China's Innovation Policy Is a Wake-Up Call for America* (2011); *A New Geography of Knowledge in the Electronics Industry? Asia's Role in Global Innovation Networks* (2009); *Can Chinese IT Firms Develop Innovative Capabilities within Global Knowledge Networks?* (2008); *China's Emerging Industrial Economy-Insights from the IT Industry* (with Barry Naughton) (2007); *Innovation Offshoring-Asia's Emerging Role in Global Innovation Networks* (2006); "Complexity and Internationalization of Innovation: Why is Chip Design Moving to Asia?", *International Journal of Innovation Management*, 2005; "Limits to Modularity: Reflections on Recent Developments in Chip Design", *Industry and Innovation*, 2005; *International Production Networks in Asia: Rivalry or Riches?* (2000); and *Technological Capabilities and Export Success: Lessons from East Asia* (1998).

**MARK D. HILL** is a professor of computer science and electrical and computer engineering at the University of Wisconsin–Madison. Dr. Hill's research targets computer design and evaluation. He has made contributions to parallel computer system design (e.g., memory-consistency models and cache coherence), memory-system design (caches and translation buffers), computer simulation (parallel systems and memory systems), software (e.g., page tables and cache-conscious optimizations for databases and pointer-based codes), and transactional memory. For example, he is the inventor of the widely used 3C model of cache behavior (compulsory, capacity, and conflict misses).

Dr. Hill's current research is mostly part of the Wisconsin Multifacet Project that seeks to improve the multiprocessor servers that form the computational infrastructure for Internet Web servers, databases, and other demanding applications. The Multifacet work focuses on using the transistor bounty provided by Moore's Law to improve multiprocessor performance, cost, and fault tolerance, while also making these systems easier to design and program.

Dr. Hill was named an ACM fellow (2004) for contributions to memory consistency models and memory system design, elevated to a fellow of the IEEE (2000) for contributions to cache memory design and analysis, and was awarded the ACM SIGARCH Distinguished Service Award in 2009. He has won three important University of Wisconsin awards: Kellett Mid-Career in 2010, Vilas Associate in 2006, and Romnes Faculty Fellowship in 1997. He co-edited *Readings in Computer Architecture* in 2000, is coinventor of more than 30 U.S. patents (several of which have been coissued in the European Union and Japan), was an ACM SIGARCH director (1993–2007), and won a National Science Foundation Presidential Young Investigator award in 1989. He is coauthor of five papers selected by IEEE *Micro* Top Picks and co-won the best paper award at the *International Conference on Very Large Databases* (VLDB) in 2001. He has held visiting positions at Advanced Micro Devices (2011), University of Washington (2011), Columbia University (2010), Polytechnic University of Catalonia (2002–2003) and Sun Microsystems (1995–1996). Dr. Hill earned a Ph.D. in computer science from the University of California, Berkeley, in 1987, an M.S. in computer science from UC Berkeley in 1983, and a B.S.E. in computer engineering from the University of Michigan–Ann Arbor in 1981.

**STEPHEN W. KECKLER** is the senior director of architecture research at NVIDIA and professor of both computer science and electrical and computer engineering at the University of Texas (UT) at Austin, where he has served on the faculty since 1998. His research interests include parallel computer architecture, technology-scalable architectures, very-large-scale integration (VLSI) design, high-performance computing, energy-efficient computing, and on-chip interconnection networks. He has developed both commercial chips at Intel and parallel computing prototype chips at MIT and UT Austin. At MIT, he was the principal architect of the M-Machine multicomputer, a research machine that was one of the first multicore processors and included extremely efficient inter-thread communication and synchronization mechanisms. His research team at UT Austin developed scalable parallel processor and memory system architectures, including nonuniform cache architectures; explicit data graph execution processors, which merge dataflow execution with sequential memory semantics; and micro-interconnection networks to implement distributed processor protocols. His research team at NVIDIA is developing extreme energy-efficient computing technologies for massively parallel chips and systems.

Dr. Keckler was named fellow of the ACM (2011) for contributions to computer architectures and technology modeling, and was elevated to a fellow of the IEEE for contributions to computer architectures and memory systems. He received the 2003 ACM Grace Murray Hopper Award for ground-breaking analysis of technology scaling for high-performance processors that sheds new light on the methods required to maintain performance improvement trends in computer architecture, and on the design implications for future high-performance processors and systems. He won an NSF CAREER award (2000), was selected as an Alfred P. Sloan research fellow (2002), won the Edith and Peter O'Donnell Award for Engineering (2010), and won six IBM Faculty Partnership awards. Dr. Keckler is coauthor of four papers selected by IEEE *Micro* Top Picks and co-won best paper awards at the 2009 *International Symposium on Architectural Support for Programming Languages and Operating Systems* (ASPLOS) and the 2011 *International Symposium on Performance Analysis of Systems and Software* (ISPASS). He has also won top teaching awards at the University of Texas at Austin, including the College of Natural Sciences Teaching Excellence Award (2001) and the President's Associates Teaching Excellence Award (2007). Dr. Keckler also served as a member of the Defense Science Study Group, sponsored by the Defense Advanced Research Projects Agency (DARPA), (2008–2009). Dr. Keckler earned a B.S. in electrical engineering from Stanford University (1990), and an M.S. (1992) and a Ph.D. (1998) in computer science from the Massachusetts Institute of Technology.

**DAVID LIDDLE** has been a partner at U.S. Venture Partners, a Silicon Valley–based venture capital firm, since 2000. He cofounded, and between 1992 and 1999, he served as president and CEO of, Interval Research Corporation, a Silicon Valley–based laboratory and incubator for new businesses focusing on broadband, consumer devices, interaction design, and advanced technologies. Prior to cofounding Interval with Paul Allen, Dr. Liddle founded Metaphor, which was acquired by IBM in 1991, which named him vice president of business development for IBM Personal Systems. Dr. Liddle's extensive experience in research and development has focused largely on human-computer interactions and includes 10 years at Xerox Palo Alto Research Center (PARC), from 1972 to 1982. He has been a director of Sybase, Broderbund Software, Borland International, and Ticketmaster, and is currently on the board of the New York Times Company. His board involvement at U.S. Venture Partners includes Electric Cloud, Instantis, Klocwork, MaxLinear, and Optichron. Dr. Liddle has served on the DARPA Information Science and Technology Committee, and as chair of the National Academy of Sciences Computer Science and Telecommunications Board.

Dr. Liddle earned a B.S. in electrical engineering at the University of Michigan and a Ph.D. in electrical and engineering and computer science at the University of Toledo, where his dissertation focused on reconfigurable computing machines and theories of encryption, encoding, and signal recovery. His contributions to human-computer interaction design earned him the distinction of senior fellow at the Royal College of Art.

**KATHRYN MCKINLEY** is principal researcher at Microsoft Research and holds an Endowed Professorship of Computer Science at the University of Texas at Austin. She previously was a professor at the University of Massachusetts, Amherst.

Dr. McKinley's research interests include programming language implementation, compilers, memory management, runtime systems, security, reliability, and architecture. Her research group has produced numerous tools, algorithms, and methodologies that are in wide research and industrial use, such as the DaCapo Java Benchmarks, the TRIPS compiler, the Hoard memory manager, the Memory Management Toolkit (MMTk), and the Immix mark-region garbage collector. For example, the Apple operating system uses the Hoard memory management algorithm, the TRIPS compiler was the first demonstration of a compiling general-purpose programming language to execute on a dataflow architecture, and the DaCapo Benchmarks are the most widely used Java benchmarks for performance and verification in both research and testing.

Dr. McKinley was named an ACM fellow (2008) for contributions to compilers and memory management and an IEEE fellow (2011) for contributions to compiler technologies. She was awarded the 2011 ACM SIGPLAN Distinguished Service Award. She has served as the technical program chair for ASPLOS, PACT, PLDI, ISMM, and CGO (ACM and IEEE conferences). She was coeditor-in-chief of ACM's *Transactions on Programming Language Systems* (or TOPLAS) (2007–2010). She was a Computer Research Association's Committee on the Status of Women in Computer Science (CRA-W) board member (2009–2011) and is currently a cochair of CRA-W (2011–present), which seeks to improve the participation of women in computing research nationwide. Her research awards include two CACM Research Highlights Invited Papers (2012, 2008), IEEE *Micro* Top Picks (2012), Best Paper at ASPLOS (2009), David Bruton Jr. Centennial Fellowship (2005–2006), six IBM Faculty Fellowship Awards (2003–2008), and an NSF CAREER Award (1996–2000). She is a recipient of the 2011 ACM SIGPLAN Software Award. Dr. McKinley has graduated fourteen Ph.D. students. She received a B.A., M.S., and Ph.D. from Rice University.

# B

# Identifying Hubs of Research Activity in Key Areas of S&T Critical to this Study

As a data-gathering experiment, the committee solicited input from experts on their sense of where innovation and engagement are taking place related to the power and performance challenges for sustaining growth in computing performance. In particular, the committee asked experts to identify leading researchers around the globe focused on the challenges of sustaining growth in computer performance in (1) semiconductor device scaling, (2) power efficiency in computing hardware, (3) parallel programming and models to leverage multicore and other novel architecture, (4) chip architectures, and (5) runtime and software infrastructure for power-efficient and scalable computing.

Approximately 170 leading researchers were identified, based on input solicited from a dozen computer scientists, engineers, and recommendations by the committee.[1] Approximately three-quarters of those identified were based in the United States. Individuals who were identified by at least three people were deemed to be "hubs" of concentrated research activity for the purposes of the committee's analysis. Publication data (for the years 2001–2011) for each of these individuals, or hubs, was collected using SciVerse Scopus[2] (resulting in a total number of 1,081 publications and 5,685 authors, 1,368 of which are unique authors).

Using this publication data, the committee generated a coauthor publication network map that includes all identified hubs of research activity in semiconductor scaling, architecture, and parallel programming.[3,4] A visualization of the publication network for these advanced research areas critical to the computing performance challenge is shown in Figure B-1. The goal of this exercise is not to highlight individually-identified researchers, but rather to present a methodology that allows gleaning, at least in a rough, qualitative sense, of potential insights from the connectedness between the hubs of research activity, as well as between U.S. and international research communities.

Figure B-1 shows a map of highly connected circles. Each circle represents an individual researcher, and each line between two circles represents a coauthored publication. The size[5] of each circle corresponds approximately to the total number of papers that researcher has authored, and the width of each line corresponds to the number of coauthored papers shared between two researchers. Hubs of research activity are colored in yellow and labeled with letters corresponding to their institutional affiliation.[6] All other circles are

---

[1] The following noncommittee members contributed to this data-collection exercise: Alex Aiken (Stanford University), Mark Bohr (Intel), Robert Colwell (DARPA), Bob Doering (Texas Instruments), Bryan Ford (Yale University), David Patterson (University of California-Berkeley); David Srolovitz (A*STAR), and Dennis Sylvester (University of Michigan).

[2] See www.scopus.com. Last accessed on August 11, 2012.

[3] Coauthor publication network maps are not shown for advanced research in power efficiency in computing hardware or in runtime and software infrastructure for power-efficient and scalable computing due to limited overlap in researcher nominations.

[4] Coauthor publication network maps were generated using the Science of Science (Sci²) Tool, available at http://sci2.cns.iu.edu.

[5] The size of each node is calculated as a fraction of the largest number of papers authored and/or coauthored by a single individual.

[6] Affiliations associated with each hotspot are as follows: (A) The University of California, Berkeley; (B) Massachusetts Institute of Technology; (C) Stanford University; (D) University of Illinois at Urbana-Champaign; and (E) Advanced Micro Devices, Inc.

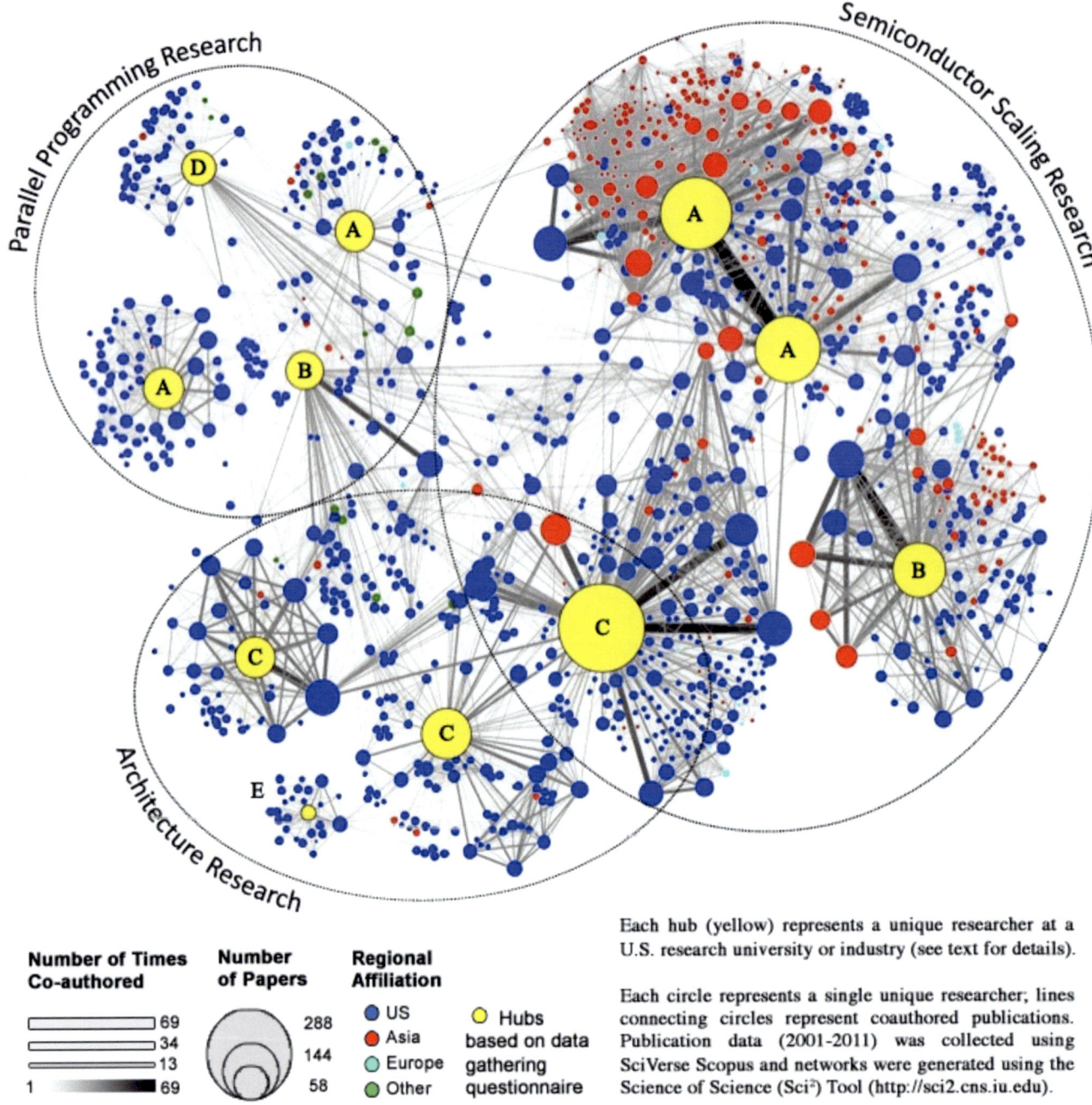

FIGURE B-1 Coauthorship networks of hubs of research activity in three areas of advanced computing research.

colored according to the geographic location of that researcher, as indicated in the figure.[7] Three large circles marked by dotted lines are used to bin the hubs of research activity by each of their associated areas of research (e.g., semiconductor scaling, architecture, and parallel programming).

It is worth noting that this mapping approach differs from traditional bibliometric analyses of coauthored publication data because the primary nodes (hubs) examined were identified based on the committee's data input solicitations (as opposed to selecting hubs on the basis of total number of publications or of most cited

---

[7] Color coding by region (e.g., U.S., Asia, Europe, Other) was determined by using each authors' geographic location listed on his or her conference publication. Addresses were not mapped using Sci[2] software.

publications). While the sampling size of the data solicitations is small, Figure B-1 reveals several interesting features that may be useful for subsequent analysis.

For example, all hubs are located in the United States, and all but one of these hubs are U.S. research universities. The coauthorship network maps show that some areas of research (and some researchers specifically) tend to collaborate on a nation- or region-specific basis or both. For example, chip architecture and parallel programming networks are primarily U.S. based with limited participation by Europe and Asia. In contrast, semiconductor device-scaling networks show a significant number of collaborations with Asia. In particular, Taiwan holds the vast majority share of Asia's representation, followed by Singapore and Japan (data not shown).

While the individual hubs do not generally show a significant degree of connectivity with one another (with exception to two hubs in the semiconductor scaling networks), the semiconductor scaling and chip architecture networks appear to be highly interconnected. In fact, both of these networks share a common hub. In contrast, researchers within the parallel programming networks display much less connectivity.

Increasing circle size, increasing connectedness between researchers, and widening lines between researchers may all be useful indicators for identifying emerging hubs of research activity. For instance, a small circle with many connections might suggest an individual who publishes less but collaborates frequently and is thereby more intimately connected to the global knowledge network. In addition, a wide line between a small circle and an established research hub might suggest a promising early-career researcher who hails from a strong research lineage. This analysis could also be extended by observing how coauthored publication networks change over time.

In summary, this methodology presents a unique approach for identifying emerging, as well as established, hubs of research activity in three areas of science and technology. However, given the small sampling size of the data solicitations, this experiment is not intended to provide any assessment or interpretation of the hubs themselves (or of trends apparent in the network maps). Rather, the goal of this experiment is to demonstrate an approach that could be extended and/or modified (e.g., to include statistically valid data-gathering methodologies) for subsequent in-depth exploration in any number of research areas.

# C

# Contributors to the Study

Although the briefers listed below provided much useful information of various kinds to the Committee on Global Approaches to Advanced Computing, they were not asked to endorse the content of this study, nor did they see the final draft of this report before its release.

**BRIEFERS AND PRESENTERS TO THE COMMITTEE**
**SEPTEMBER 19–20, 2011**
**WASHINGTON, DC**

Connie Chang, Independent Consultant
Krisztian Flautner, ARM Limited
David Jakubek, Department of Defense
Luc Soete, United Nations University – Maastricht Economic and Social Research Institute on Innovation and Technology (UNU-MERIT)
Gregory Tassey, National Institute of Standards and Technology

# D

# Findings and Recommendations from *The Future of Computing Performance: Game Over or Next Level?*

The following findings and recommendations are repeated from the National Research Council's report, The Future of Computing Performance: Game Over or Next Level?[1]

Findings:

- The information technology sector itself and most other sectors of society—for example, manufacturing, financial and other services, science, engineering, education, defense and other government services, and entertainment—have grown dependent on continued growth in computing performance.
- After many decades of dramatic exponential growth, single processor performance is increasing at a much lower rate, and this situation is not expected to improve in the foreseeable future.
- The growth in the performance of computing systems—even if they are multiple-processor parallel systems—will become limited by power consumption within a decade.
- There is no known alternative to parallel systems for sustaining growth in computing performance; however, no compelling programming paradigms for general parallel systems have yet emerged.

Recommendations:

- Invest in research in and development of algorithms that can exploit parallel processing.
- Invest in research in and development of programming methods that will enable efficient use of parallel systems not only by parallel-systems experts but also by typical programmers.
- Focus long-term efforts on rethinking of the canonical computing "stack"—applications, programming language, compiler, runtime, virtual machine, operating system, hypervisor, and architecture—in light of parallelism and resource-management challenges.
- Invest in research on and development of parallel architectures driven by applications, including enhancements of chip multiprocessor systems and conventional data-parallel architectures, cost-effective designs for application-specific architectures, and support for radically different approaches.
- Invest in research and development to make computer systems more power-efficient at all levels of the system, including software, application-specific approaches, and alternative devices. Such efforts should address ways in which software and system architectures can improve power efficiency, such as by exploiting locality and the use of domain-specific execution units. R&D should also be aimed at making logic gates more power-efficient. Such efforts should address alternative physical devices beyond incremental improvements in today's CMOS circuits.
- To promote cooperation and innovation by sharing, encourage development of open interface standards for parallel programming rather than proliferating proprietary programming environments.
- Invest in the development of tools and methods to transform legacy applications to parallel systems.
- Incorporate in computer science education an increased emphasis on parallelism, and use a variety of methods and approaches to better prepare students for the types of computing resources that they will encounter in their careers.

---

[1] NRC, 2011, *The Future of Computing Performance: Game Over or Next Level?*, Washington, D.C.: The National Academies Press (available online at http://www.nap.edu/catalog.php?record_id=12980).

# E

# Dennard Scaling and Implications

The following description was taken from the 2010 National Research Council Computer Science and Telecommunications Board (CSTB) report *The Future of Computing Performance: Game Over or Next Level?*[1]

"In a classic 1974 paper, reprinted in Appendix D, Robert Dennard et al. showed that the MOS transistor has a set of very convenient scaling properties.[10] The scaling properties are shown in Table 3.1, taken from that paper. If all the voltages in a MOS device are scaled down with the physical dimensions, the operation of the device scales in a particularly favorable way. The gates clearly become smaller because linear dimensions are scaled. That scaling also causes gates to become faster with lower energy per transition. If all dimensions and voltages are scaled by the scaling factor κ (κ has typically been 1.4), after scaling the gates become $(1/\kappa)^2$ their previous size, and $\kappa^2$ more gates can be placed on a chip of roughly the same size and cost as before. The delay of the gate also decreases by $1/\kappa$, and, most importantly, the energy dissipated each time the gate switches decreases by $(1/\kappa)^3$. To understand why the energy drops so rapidly, note that the energy that the gate dissipates is proportional to the energy that is stored at the output of the gate. That energy is proportional to a quantity called capacitance[11] and the square of the supply voltage. The load capacitance of the wiring decreases by $1/\kappa$ because the smaller gates make all the wires shorter and capacitance is proportional to length. Therefore, the power requirements per unit of space on the chip ($mm^2$), or energy per second per $mm^2$, remain constant:

Power = (number of gates)($C_{Load/gate}$)(Clock Rate)($V_{supply}^2$)
Power density = $N_g C_{load} F_{clk} V_{dd}^2$
 $N_g$ = CMOS gates per unit area
 $C_{load}$ = capacitive load per CMOS gate
 $F_{clk}$ = clock frequency
 $V_{dd}$ = supply voltage
Power density = $(\kappa^2)(1/\kappa)(\kappa)(1/\kappa)^2 = 1$

That the power density (power requirements per unit space on the chip, even when each unit space contains many, many more gates) can remain constant across generations of CMOS scaling has been a critical property underlying progress in microprocessors and in ICs in general. In every technology generation, ICs can double in complexity and increase in clock frequency while consuming the same power and not increasing in cost. Given that description of classic CMOS scaling, one would expect the power of processors to have remained constant since the CMOS transition, but this has not been the case. During the late 1980s and early 1990s, supply voltages were stuck at 5 V for system reasons. So power density would have been expected to increase as technology scaled from 2 mm to 0.5 mm. However, until recently supply voltage has scaled with technology, but power densities continued to increase."

---

[1] NRC, 2011, *The Future of Computing Performance: Game Over or Next Level?*, Washington, D.C.: The National Academies Press (available online at http://www.nap.edu/catalog.php?record_id=12980).

# F

# Pilot Study of Papers at Top Technical Conferences in Advanced Computing

The committee found it challenging to identify reliable and robust nontraditional indicators for assessing a nation's technological research capabilities specific to the computing performance challenges outlined in Chapter 1. After significant methodological consideration and debate, the committee conducted a pilot study to determine whether a bibliometric analysis of papers at select prestigious conferences in advanced computing could provide a useful snapshot of a nation's capabilities in specific technology areas critical to the study's charge.

The committee noted the strengths and challenges of a methodology that is both objective (e.g., conference publication data) and subjective (e.g., using committee expertise to identify specific conferences for analysis). By excluding traditional journal publications—as well as papers from conferences not considered to be representative of the most relevant and leading research—some relevant research may have been excluded; alternative samplings of conferences could also yield different results. On the other hand, a selective sampling of conferences may better support a more focused assessment of research efforts across specific technology areas (i.e., semiconductors and nanoscale devices and circuits, architecture, programming systems, and applications).

An ideal analysis would include all papers relevant to the computing performance challenges outlined in Chapter 1, whether published at conferences or in traditional journals, weighted by citations and impact factors, as well as expert judgment.[1] Given limitations in time and resources, however, this was simply not feasible. Further, the committee's analysis represents only one of likely many different and potentially useful measures of a nation's research capabilities and innovativeness. These factors considered, the committee's methodological approach and insights from its assessment are presented here as a pilot effort that will, no doubt, benefit from deeper, subsequent exploration by others.

## F.1 Methodological Overview of Conference Paper Authorship Analysis

To assess a nations' technology-specific research capabilities, the committee analyzed paper authorship—specifically, the geographical locations of authors—in many of the top technical conferences in four research areas most closely related to the technological challenges outlined in Chapter 1. These relate to the computing performance challenge and the shift to multicore processors: semiconductor devices and circuits, computer architecture, programming systems, and applications.

*F.1.1 Rationale for Conferences as a Preferred Venue*

In the computing community—unlike many other science and engineering disciplines—conference papers are often the publication venue of choice rather than journals.[2] In fact, highly selective computer science

---

[1] Council of Canadian Academies, 2012, *Informing Research Choices: Indicators and Judgment – The Report of the Expert Panel on Science Performance and Research Funding* (available at http://www.scienceadvice.ca/uploads/eng/assessments% 20and%20publications%20and%20news%20releases/Science%20 performance/SciencePerformance_FullReport_EN_Web.pdf).

[2] David Patterson, Lawrence Snyder, and Jeffrey Ullman, 1999, "Best Practices Memo: Evaluating Computer Scientists and Engineers for Promotion and Tenure" in *Computing Research News*, June. Available at http://www.cra.org/uploads/documents/resources/bpmemos/tenure_review.pdf. Last accessed on August 15, 2012.

conferences often have higher citation indices and greater venue impact compared to related computing journals.[3,4] The importance of conference publications within this community is consistent with a 2011 NRC report, *A Data-Based Assessment of Research-Doctorate Programs in the United States*,[5] which indicates that "for the field of computer science, refereed conference papers are an important form of scholarship."

In this report, conferences are the preferred venue over journals for the computer science-focused areas described in Sections F.1.3–F.1.6 for several reasons. First, these conferences tend to have a much shorter time from submission to publication than journals in the area, resulting in the most recent, significant innovations appearing at the conferences first.[6] Second, conferences provide a larger sample size than journals of highly recognized recent top-quality research. Third, the conferences identified by the committee often have more focused research interests compared to journals that would publish related (albeit less recent and possibly less regarded), but broader works.[7]

For example, in advanced architecture research, relevant to the challenges described in Chapter 1, architecture papers appear in a wider range of journals that include more than just computer architecture (e.g., *IEEE Transactions on Computers*, *ACM Transactions on Computer Systems*, and *IEEE Transactions on Parallel and Distributed Systems*). Achieving the same level of topical focus would require disaggregation of the journal data on a paper-by-paper basis. This is simply not feasible for the wide range of conferences selected and papers analyzed for this report. Thus, analysis of conference papers allows a more focused assessment of the research areas identified by the committee as critical for sustaining computing performance and the shift to multicore processors, as opposed to computer science generally.

Similar to prestigious journals, premier conferences in each of the targeted hardware, architecture, and software research areas are highly competitive and conference submissions are rigorously peer-reviewed. Conferences are also competitive publishing venues because representation is professionally beneficial. Conferences provide an opportunity for researchers to share new research, to learn from others, and to gain exposure to recent and significant research efforts.[8,9]

*F.1.2 Determination of a Nation's Paper Contributions at Conferences*

As a proxy indicator for a nation's technology-specific research capabilities, the committee analyzed the weighted distribution of authors for research papers given at the conferences listed and described in Sections F.1.3–F.1.6. To do this, the committee noted the home nation for each author (defined as the geographic location of that author's affiliation listed on each paper's title page) and computed each nation's weighted authorship contribution to each conference paper. It is important to note that this analysis does not distinguish between a U.S.-based author who is a U.S. citizen and a U.S.-based author who is not a U.S. citizen (and may eventually return to his or her home country), which may, in some cases, diminish a nation's research representation at the sampled conferences.[10]

---

[3] It is worth noting that industry participation in these conferences may be limited by the demise of central research labs, less emphasis on outside presentations and publications, and reluctance to report on the most important research that companies are performing.

[4] As one example, CiteSeer, which keeps statistics about computer science publications, reports a venue impact of 0.14, 0.08, and 0.6 for three architecture research conferences analyzed in this report (*High-Performance Computer Architecture*, *International Symposium on Computer Architecture*, and the *International Symposium on Microarchitecture*, respectively) compared to 0.02 and 0.01 for two related architecture research journals: *IEEE Transactions on Computers* and *IEEE Transactions on Parallel and Distributed Systems*, respectively. See http://citeseer.ist.psu.edu/stats/venues. Last accessed on August 15, 2012.

[5] NRC, 2011, *A Data-Based Assessment of Research-Doctorate Programs in the United States*, Washington, D.C.: The National Academies Press (available online at https://download.nap.edu/rdp/index.html?)

[6] This is particularly relevant to the analysis of recent conference papers in Appendix F.2.

[7] One exception of a journal that is equally regarded with a similarly themed premier conference is the *IEEE Journal of Solid-State Circuits* (JSSC). JSSC publishes approximately 200 papers per year, with the papers typically being 10-15 pages long. By comparison, the *International Solid-State Circuits Conference (ISSCC)* typically publishes the same number of papers, but each paper is 3 pages long. As both venues are highly regarded and may represent a similar sample of annual papers, a more in-depth analysis of contributions in this area would benefit by addressing both ISSCC and JSSC.

[8] David Patterson, Lawrence Snyder, and Jeffrey Ullman, 1999, "Best Practices Memo: Evaluating Computer Scientists and Engineers for Promotion and Tenure" in *Computing Research News*, June. Available at http://www.cra.org/uploads/documents/resources/bpmemos/tenure_review.pdf. Last accessed on August 15, 2012.

[9] NRC, 1994, *Academic Careers for Experimental Computer Scientists and Engineers*, Washington, D.C.: The National Academies Press (available online at http://www.nap.edu/openbook.php?record_id=2236).

[10] 2010 National Science Foundation data show that the share of non-U.S. citizens receiving U.S. doctoral degrees in natural sci-

A nation's weighted[11] contribution to each conference paper is merely the number of authors geographically located in (or professionally affiliated with) that particular nation divided by the total number of authors of the paper. Thus, the weighted contributions for a given paper always sum to one. To compute the weighted percentage of papers contributed by a nation at a given conference, each nation's weighted contributions to each conference paper are summed and then divided by the total number of papers given at that conference. The committee believes this measurement—as opposed to total papers or total authors—better reflects a nation's authorship contribution.[12] This measure also has the distinct advantage that it is public information[13] and can be extended and reproduced by others.

Based on the technical challenges outlined in Chapter 1, the committee identified four research areas critical to addressing the computing performance challenge and the shift to multicore processors: semiconductor devices and circuits, computer architecture, programming systems, and applications. To compute the weighted percentage of papers contributed by a nation in each of these research areas, each nation's weighted contributions to each conference paper are summed across all conferences assigned to a particular research area (discussed in Sections F.1.3–F.1.6) and then divided by the total number of papers given at those conferences.[14] All but three of the conferences analyzed by the committee are sponsored or published by either (or both) the Institute of Electrical and Electronics Engineers (IEEE) or the Association for Computing Machinery (ACM), the two preeminent international technical societies in electrical engineering and computing. Based on committee members' opinions and knowledge of the fields, the following sections identify a limited number of top conferences that make available new and interesting research germane to the study's charge.

*F.1.3 Semiconductor Devices and Circuits Conferences*

As described in Chapter 1, the end of Dennard scaling has placed greater pressure on innovative devices and circuits to deliver more energy-efficient technologies for building microprocessors. To explore the research capabilities in these areas, the committee analyzed papers from three conferences—two in semiconductor and nanoscale devices and one in semiconductor circuits following the methodology described earlier in Section F.1.2. These conferences are described below.

- *International Electron Devices Meeting* (IEDM). As stated on the conference Web site, IEDM is "the world's pre-eminent forum for reporting technological breakthroughs in the areas of semiconductor and electronic device technology, design, manufacturing, physics, and modeling. IEDM is the flagship conference for nanometer-scale CMOS (complementary-symmetry metal-oxide-semiconductor) transistor technology, advanced memory, displays, sensors, MEMS (microelectromechanical systems) devices, novel quantum and nano-scale devices and phenomenology, optoelectronics, devices for power and energy harvesting, high-speed devices, as well as process technology and device modeling and simulation. The conference scope not only encompasses devices in silicon, compound and organic semiconductors, but also in emerging material systems. IEDM is truly an international conference, with strong representation from speakers from around the globe."[15] In 2011, IEDM included 36 sessions encompassing more than 200 papers. IEDM is sponsored by the IEEE.

---

ences and engineering is high and increasing at a higher rate than for U.S. citizens (available at http://www.nsf.gov/statistics/seind10/pdf/overview.pdf; last accessed on September 2, 2012). Additionally, a 2007 report by the Oak Ridge Institute for Science and Education, *Stay Rates of Foreign Doctorate Recipients from U.S. Universities* (available at http://orise.orau.gov/files/sep/stay-rates-foreign-doctorate-recipients-2007.pdf) found that two-thirds of foreign citizens who received science or engineering doctorates from U.S. universities in 2005 continued to live in the United States in 2007.

[11]In an unweighted analysis, two countries will receive 50 percent of a particular paper's contribution to the conference, even if nine coauthors are located in one country and one coauthor is located in the other country.

[12]Examination of the country of origin of members of conference programming committees may be a useful future activity and provide additional insight for assessing a nation's interest and capabilities in a particular technological field.

[13]While bibliometric databases, such as SciVerse Scopus and Web of Science, provide some conference publication and citation data, comprehensive and consistent data for each conference across the time periods analyzed in the report do not exist. For example, Scopus includes only limited or no coverage of ECOOP, Eurographics, OSDI, SC, SOSP, VLDB, and WWW, and large gaps in annual coverage exist for ISCA, MICRO, POPL, and PPoPP.

[14]This allows all amassed papers in each research area to be weighted equally. In contrast, by first calculating a nation's weighted percentage of contributed papers for each conference (as reported in Appendix G) and then averaging across all conferences assigned to a particular research area, a bias is introduced that could skew the overall average in favor of those conferences with larger numbers of presented papers.

[15]See http://www.his.com/~iedm/. Last accessed on January 9, 2012.

- *International Conference on Nanotechnology* (NANO). From the conference Web site: "NANO is the flagship IEEE conference in Nanotechnology, which makes it a must for students, educators, researchers, scientists and engineers alike, working at the interface of nanotechnology and the many fields of electronic materials, photonics, bio- and medical devices, alternative energy, environmental protection, and multiple areas of current and future electrical and electronic applications. In each of these areas, NANO is the conference where practitioners will see nanotechnologies at work in both their own and related fields, from basic research and theory to industrial applications."[16] In 2011, NANO included more than 400 papers. NANO is sponsored by the IEEE.
- *International Solid-State Circuits Conference* (ISSCC). As stated on the conference Web site, ISSCC is "the premier forum for the presentation of advances in solid-state circuits and systems-on-a-chip."[17] ISSCC topics include advanced memory circuits, low-power circuits, high-speed signaling, and microprocessors, among many others. In 2011, ISSCC included 28 sessions encompassing more than 200 papers. ISSCC is sponsored by the IEEE.

*F.1.4 Computer Architecture Conferences*

Computer architecture includes the design and study of computer hardware implementations and computer design at the hardware-software boundary. Computer architects seek to make computers faster, lower power, cheaper, more reliable, and easier to program. Many computer architecture researchers focus on parallel and multicore systems. The committee analyzed conference papers from four top-flight conferences, described below.

- *International Symposium on Architectural Support for Programming Languages and Operating Systems* (ASPLOS). ASPLOS "is the premier forum for multidisciplinary systems research, spanning hardware, computer architecture, compilers, languages, operating systems, networking, and applications,"[18] and includes papers on parallel hardware and software. In 2011, ASPLOS included 14 sessions with 32 papers. ASPLOS is sponsored by the ACM.
- *International Symposium on High Performance Computer Architecture* (HPCA). HPCA covers many of the same topics as ISCA and MICRO.[19] In 2011, HPCA included 14 sessions and 46 papers. HPCA is sponsored by the IEEE.
- *International Symposium on Computer Architecture* (ISCA). As stated on the conference Web site, ISCA is "the premier forum for new ideas and experimental results in computer architecture,"[20] including parallel architecture and multicore systems. In 2011, ISCA included 14 sessions with 40 papers. ISCA is sponsored by ACM and the IEEE.
- *International Symposium on Microarchitecture* (MICRO). As stated on the conference Web site, MICRO "brings together researchers in fields related to microarchitecture, compilers, chips, and systems for technical exchange on traditional microarchitecture topics and emerging research areas."[21] In 2011, MICRO included 13 sessions with 44 papers. MICRO is sponsored by ACM and the IEEE.

Roughly, the same community of researchers publishes in and attends the conferences described above, although ASPLOS includes additional research areas on the boundary between computer architecture, programming languages, and operating systems.

*F.1.5 Programming Systems Conferences*

In this report, the committee focuses on (1) programming systems that encompass programming language design and implementation, and (2) programming tools, including programming models, languages, compilers, runtime systems, and virtual machines for parallel systems that are necessary to enable applications to exploit emerging silicon trends and chip architectures. The following five top conferences were analyzed, which collectively cover a range of programming system technologies:

- *European Conference on Object-Oriented Programming (ECOOP)*. ECOOP covers "topics on object-oriented technologies, software development, systems, languages and

---

[16] See http://ieeenano2011.org/. Last accessed on January 9, 2012.
[17] See http://isscc.org/. Last accessed on January 9, 2012.
[18] See research.microsoft.com/en-us/um/Cambridge/events/asplos_2012. Last accessed on January 9, 2012.
[19] See www.ece.lsu.edu/hpca-18/. Last accessed on January 9, 2012.
[20] See Isca2012.ittc.ku.edu. Last accessed on January 9, 2012.
[21] See www.microarch.org/micro44/. Last accessed on January 9, 2012.

applications."[22] ECOOP was established in 1987. ECOOP and OOPSLA are peer conferences, were established within 1 year of each other, and have followed the same historical trends on topics. Whereas many of the other conferences the committee sampled are often hosted in the United States or Canada, ECOOP has only ventured outside of Europe twice (1990, 2012). In 2011, ECOOP included 9 sessions and 26 papers. Since 2007, ECOOP has been sponsored by ACM.

- *Object-Oriented Programming, Systems, Languages, and Applications (OOPSLA).* OOPSLA "embraces all aspects of software construction and delivery" and is "a premier forum for software innovation."[23] OOPSLA started in 1986 when object-oriented programming systems were emerging to be a forum for researchers and practitioners to explore this new paradigm. Object-oriented programming subsequently became a dominant paradigm. Now OOPSLA is much broader and covers the same topics as PLDI and POPL. In 2011, OOPSLA included 17 sessions with 61 papers (the most in its history). OOPSLA is sponsored by the ACM.

- *Programming Language Design and Implementation (PLDI).* PLDI focuses "on the design, implementation, development, and use of programming languages. [It] emphasizes innovative and creative approaches to compile-time and runtime technology; novel language designs and features; and results from implementations."[24] Parallel programming systems are a significant component of PLDI. In 2011, PLDI included 20 sessions with 55 papers. PLDI is sponsored by the ACM.

- *Symposium on Principles of Programming Languages (POPL).* POPL is the leading "forum for the discussion of all aspects of programming languages and systems, with emphasis on how principles underpin practice."[25] POPL includes research on the principles of parallel programming systems. In 2011, POPL included 16 sessions with 49 papers. POPL is sponsored by the ACM.

- *Symposium on Principles and Practice of Parallel Programming (PPoPP).* "PPoPP is a forum for leading work on all aspects of parallel programming, including foundational and theoretical aspects, techniques, tools, and practical experiences."[26] Conference topics include work on concurrent and parallel (e.g., multicore, heterogeneous, and distributed) systems. In 2011, PPoPP included 8 sessions with 26 papers. PPoPP is sponsored by the ACM.

*F.1.6 Applications Conferences*

Computer applications are a tremendously broad area encompassing topics such as scientific computing, security, distributed and cloud computing, databases, and artificial intelligence. Representing the full range of these areas is beyond the scope of this report. However, since parallel application development is central to the success of multicore systems, the committee sampled seven conferences that have a strong focus on computational application needs and historically have depended on parallel and multicore systems, described below.

- *Annual Conference of the European Association for Computer Graphics (Eurographics).* Eurographics is a "Europe-wide professional Computer Graphics Association. . . that supports its members in advancing the state of the art in Computer Graphics and related fields such as Multimedia, Scientific Visualization, and Human Computer Interfaces."[27] In 2011, Eurographics included 14 sessions and 35 papers.

- *Symposium on Operating Systems Design and Implementation (OSDI).* OSDI "brings together professionals from academic and industrial backgrounds. . . [to discuss] the design, implementation, and implications of systems software."[28] In 2010, OSDI included 11 sessions with 32 papers. OSDI is sponsored by USENIX, the Advanced Computing Systems Association.

- *International Conference on Computer Graphics and Interactive Techniques (SIGGRAPH).* SIGGRAPH is the "premier international forum for disseminating new scholarly work in computer graphics and interactive techniques."[29] Graphics has a huge computational demand that has long been satisfied by parallel hardware, including both

---

[22] See http://ecoop12.cs.purdue.edu/. Last accessed on January 12, 2012.

[23] See http://researchr.org/conference/oopsla-2012. Last accessed on January 12, 2012.

[24] See pldi12.cs.purdue.edu. Last accessed on January 9, 2012.

[25] See www.cse.psu.edu/popl/12. Last accessed on January 9, 2012.

[26] See Dynopt.org/ppopp-2012/. Last accessed on January 9, 2012.

[27] See http://www.eg.org/index.php/about-eg/about-eg. Last accessed on June 29, 2012.

[28] See http://static.usenix.org/event/osdi10/cfp/. Last accessed on June 29, 2012.

[29] See www.siggraph.org/s2011/. Last accessed on January 9, 2012.

graphics processors and multicore processors. In 2011, SIGGRAPH included 28 sessions with 115 papers. SIGGRAPH is sponsored by the ACM.

- *International Conference for High Performance Computing, Networking, Storage, and Analysis (SC).* The SC conference engages "the international community in high performance computing, networking, storage, and analysis."[30] SC is the premier conference on supercomputing applications and systems, and has been a leading venue focusing on parallel systems ranging from traditional supercomputers to many-cabinet machines to multicore to systems built from multicore hardware. In 2011, SC included 74 papers. SC is sponsored by the ACM and the IEEE.

- *Symposium on Operating Systems Principles (SOSP).* The SOSP conference focuses on "research related to the design, implementation, analysis, evaluation, and deployment of computer systems software. . . [taking] a broad view of the systems area and solicits contributions from many fields of systems practice, including, but not limited to, operating systems, file and storage systems, distributed systems, mobility, security, embedded systems, dependability, system management, peer-to-peer systems, and virtualization."[31] In 2011, SOSP included 9 sessions and 28 papers. SOSP is sponsored by the ACM.

- *International Conference on Very Large Databases (VLDB).* The VLDB conference covers "current issues in data management, database and information systems research."[32] Database applications are particularly instructive in this setting, because the database community has developed and matured parallel algorithms and technologies that exploit parallel hardware. In 2011, VLDB included 30 sessions and 104 papers (18.1 percent acceptance rate). VLDB is sponsored by the nonprofit organization Very Large Data Base Endowment Inc.

- *International World Wide Web Conference (WWW).* The WWW conference "aims to provide the world a premier forum for discussion and debate about the evolution of the Web, the standardization of its associated technologies, and the impact of those technologies on society and culture."[33] The explosive growth in the numbers of these applications and their scale and parallelism make them well suited to this study. In 2011, WWW included 27 sessions and 81 papers (12.5 percent acceptance rate). Conference proceedings are published by the ACM.

In total, the analyses of conference data presented in Sections F.2 and F.3 represent aggregated results from 19 conference series (i.e., ASPLOS, ECOOP, Eurographics, HPCA, IEDM, ISCA, ISSCC, MICRO, NANO, OOPSLA, OSDI, PLDI, POPL, PPoPP, SIGGRAPH, SC, SOSP, VLDB, and WWW). Four time points (1996, 2001, 2006, and 2011)[34] were analyzed for each conference series, with exception to NANO that had only one time point, resulting in a total of 73 individual conferences comprising 4,719 papers and 23,859 authors.

*F.1.7 Global Reach and Rationale for Conferences and Other Methodological Considerations*

U.S. scientists did dominate many of the early innovations, creating the international technical societies and initiating most of the publication venues in the technology areas described in Sections F.1.3–F.1.6. The following data provide a starting point to examine the relationship between this historical U.S. advantage, as well as the location of many of the conference sites in the United States, and international research activity in these areas.

While proceedings of the selected conferences are all published in English and the conferences themselves are often held in the United States (with the exception of ECOOP, Eurographics, and WWW which were always held outside the United States for all years analyzed), more than one-third of the 73 specific conference venues analyzed by the committee were held outside the United States. As with any conference, whether held in the United States, Europe, or Asia, the location of the meeting has the potential to introduce travel biases for U.S.- and non-U.S.-based researchers. For example, insufficient travel funds or restrictive government policies could prevent qualified researchers from participating in conferences in other countries.

As one means of exploring potential travel biases due to the location of a top technical conference, the

---

[30]See www.Sc11.supercomputing.org. Last accessed on January 9, 2012.

[31]See www.sosp.org and www.sigops.org/sosp/sosp11/current. Last accessed on June 19, 2012.

[32]See www.vldb.org/2011. Last accessed on June 19, 2012.

[33]See www.www.2011india.com. Last accessed on June 19, 2012.

[34]In cases where conference proceedings authorship data was not available or where conferences were held in alternating years and thus not available for the committee's selected time points, conferences held in adjacent years were analyzed as specified in Appendix G.

committee compared regional researcher participation at 2008–2011 SIGGRAPH conferences (held in the United States) and SIGGRAPH Asia conferences, based on publication data using the methodology described in F.1.2 (see Figure F-1). While Asian participation was higher at the SIGGRAPH Asia conference than at the SIGGRAPH conference in the United States (except in 2010, when Asian participation rates were similar at both conferences), U.S. and European participation at both conferences was similar, regardless of whether the conference was held in the United States or Asia.

As an additional data point, the committee investigated the relationship between conference location and the average fraction of conference papers contributed by the United States. For all years combined, the United States contributed ~ 69 percent of papers at conferences located in the United States compared to ~62 percent when located elsewhere, as shown in Table F-1. This table also shows the U.S. share of conference papers for each of the time points analyzed in Sections F.2 and F.3, as well as the number of U.S.-located and non-U.S.-located conferences included in the committee's analysis. Table F-2 compares the average U.S. share of conference papers at U.S.- and non-U.S.-located conferences on a conference-by-conference basis. As shown in the table, U.S. paper contributions at HPCA, ISCA, MICRO, OSDI, SOSP, and VLDB, were similar or higher when the conference was located outside the United States.

These data, along with the SIGGRAPH and SIGGRAPH Asia results, suggest that for the analyses presented in this report, conference location does not induce a significant travel bias for U.S. researchers. However, as the full impact of all potential travel biases for every nation across all conferences and location cannot be measured, the above findings should be taken into consideration when interpreting the conference analyses presented in Sections F.2 and F.3.

Assessing objectively the influence of a conference is difficult because of the lack of easily comparable criteria. One starting point for this information is the Microsoft Academic Search database,[35] a publicly available Web resource that indexes publication data for some of the conferences the committee considered and mines publication and venue citations for publications and authors.[36] Table F-3 shows publication statistics, including the number of papers published and number of citations to those published papers for all of the conferences the committee selected that appear in the Microsoft database, along with a selection of European and Asian conferences that might be considered competitors. The purpose of these data is not to justify the subset of conferences identified by the committee in Sections F.1.3–F.1.6, but rather to provide an objective measure of venue impact.

In general, the papers in the international conferences the committee selected have significantly higher citations per paper than the regional conferences found in the Microsoft database.[37] The two exceptions to this are for NANO and for ECOOP. According to the Microsoft data, ECOOP has a relatively high citation rate compared with other programming system conferences, such as OOPSLA and PPoPP. The Microsoft data for OOPSLA does not reflect the expert opinion of the committee, and more careful examination of the data revealed that the number of papers reported for OOPSLA was more than twice the hand-counted technical papers.[38] This bias was corrected in the Table. CiteSeer[39] provides another objective view of programming conference impact, ranking the following venues from highest to lowest: POPL, 0.45; PLDI, 0.4; OOPSLA, 0.16; and ECOOP, 0.14. Regardless, the Microsoft data shows all the programming system venues as highly cited and, along with CiteSeer, are consistent with the committee's selection of leading technical conferences in the four research areas described in Sections F.1.3–F.1.6.

As an additional measure, attendance at the three circuits and devices conferences selected by the committee (ISSCC, IEDM, and NANO) can be compared with three notable regional conferences in these areas [IEEE *Asian Solid-State Circuits Conference* (A-SSCC), IEEE *European Solid-State Devices*

---

[35] See http://academic.research.microsoft.com/. Last accessed on June 26, 2012.

[36] In considering conference citation analysis, it is important to recognize that citation counts are influenced by several different factors, including differences in database coverage, differences in citation practices among research fields, and the age distribution of the (cited) articles.

[37] For example, the IEEE "architecture" (ISCA, MICRO, HPCA, and ASPLOS) conferences selected for analysis report 25.6–34.9 citations per paper, compared with the non-IEEE or non-ACM *European Conference on High Performance and Embedded Architecture and Compilation* (HiPEAC) (4.2 citations per paper) and *Asia-Pacific Computer Systems Architecture Conference* (ASCAC) (2.2 citations per paper).

[38] The automated system count of publications appears to have included all non-research track papers, such as poster abstracts and workshop papers, which are rarely cited.

[39] http://citeseer.ist.psu.edu/stats/venues. Last accessed on June 26, 2012.

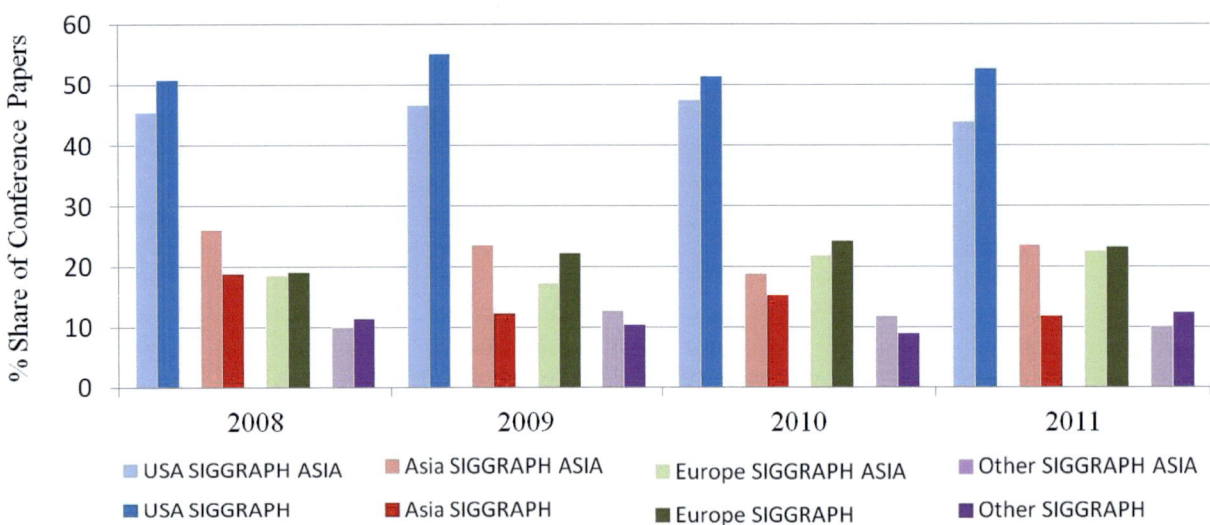

Figure F-1 Comparison of researcher participation (via publication authorship) at two conference venues: SIGGRAPH (U.S.-based) and SIGGRAPH Asia (non-U.S.-based). Data compiled from SIGGRAPH and SIGGRAPH Asia between 2008 and 2011 (~600 papers and ~1300 authors).

TABLE F-1 Comparison of U.S. Paper Contributions at U.S.-located and Non-U.S.-located Conferences

|  | Average U.S. % Share of Conference Papers | | | | |
|---|---|---|---|---|---|
|  | 1996 | 2001 | 2006 | 2011 | All years |
| U.S.-located conferences (64% of total conferences analyzed) | 71 | 66.3 | 75.3 | 64.4 | 69.2 |
| no. of conferences analyzed: | 12 | 10 | 13 | 12 |  |
| Non-U.S.-located conferences (36% of total conferences analyzed) | 58.6 | 66.8 | 53.4 | 69.6 | 62.1 |
| no. of conferences analyzed: | 6 | 8 | 5 | 7 |  |

TABLE F-2 Comparison of U.S. Paper Contributions at U.S.-located and Non-U.S.-located Conferences (By Individual Conferences)

|  | U.S. % Share of Conference Papers | | | | | | | | | |
|---|---|---|---|---|---|---|---|---|---|---|
|  | HPCA | ISCA | MICRO | OSDI | POPL | PLDI | SIGGRAPH | SOSP | VLDB | Average U.S. % Share of Papers |
| When U.S.-located | 81.8 | 88.3 | 87.3 | 90.2 | 52.1 | 63.1 | 71.4 | 96.2 | 53.8 | 76.0 |
| When located elsewhere | 83.7 | 93.1 | 86.5 | 86.8 | 44.2 | 79.0 | 52.7 | 87.7 | 58.3 | 74.7 |

*Conferences located in the United States for all years analyzed include: ASPLOS, IEDM, ISSCC, NANO, OOPSLA, PPoPP, and SC. Conferences located outside the United States for all years analyzed by the committee include: ECOOP, Eurographics, and WWW*

TABLE F-3 Conference Citation Analysis from Microsoft Scholar, June 2012

| Semiconductor Devices and Circuits | Papers | Citations | Citations/paper |
|---|---|---|---|
| ISSCC | 7,271 | 40,221 | 5.5 |
| IEDM | 9,886 | 47,925 | 4.8 |
| ESSCIRC | 1,498 | 5,090 | 3.4 |
| A-SSCC | 676 | 1,324 | 2.0 |
| ESSDERC | 1,553 | 2,511 | 1.6 |
| NANO | 2,028 | 2,266 | 1.1 |
| Architectures | | | |
| Micro | 905 | 31,607 | 34.9 |
| ASPLOS | 355 | 11,255 | 31.7 |
| HPCA | 661 | 17,891 | 27.1 |
| ISCA | 1,334 | 34,094 | 25.6 |
| HiPEAC (Europe) | 137 | 570 | 4.2 |
| ACSAC (Asia) | 339 | 750 | 2.2 |
| Programming Systems | | | |
| POPL | 1,267 | 68,945 | 54.4 |
| PLDI | 519 | 18,550 | 35.7 |
| ECOOP (Europe) | 783 | 26,054 | 33.3 |
| OOPSLA* | 835 | 14,712 | 17.9 |
| PPoPP | 319 | 5,641 | 17.7 |
| Applications | | | |
| OSDI | 255 | 20,373 | 79.9 |
| SOSP | 349 | 23,845 | 68.3 |
| VLDB | 2,739 | 12,2095 | 44.6 |
| WWW | 2,927 | 60,153 | 20.6 |
| SIGGRAPH | 3,492 | 100,567 | 28.8 |
| Eurographics (Europe) | 228 | 3,823 | 16.8 |
| Supercomputing (SC) | 2,994 | 36,868 | 12.3 |
| PARLE (Europe) | 416 | 3,804 | 9.1 |

* OOPSLA citation data was corrected by hand counting technical papers that exclude non-research track papers, which are rarely cited. Data compiled from Microsoft Scholar June 2012.

*Research Conference* (ESSDERC), and IEEE *European Solid-State Circuits Conference* (ESSCIRC)]. In 2011, ISSCC, IEDM, and NANO had significant historical attendance (according to the IEEE): 3,000, 1,500, and 400, respectfully. In contrast, 2011 attendance at A-SSCC, ESSDERC, and ESSCIRC was between 300 and 350. While conference attendance statistics do not directly correlate with quality or influence, they indicate the level of interest in the technologies and ideas found in the conference.

The committee expects that an analysis including too many (and thus a higher proportion of lower quality) conferences would underestimate the quality of leading research efforts and obscure authorship trends for leading research papers. On the other hand, too limited a sampling that overprescribes measures of conference quality and impact would also skew the assessment. Lastly, the breadth of research topics covered by a particular conference (and the same for journals that are oftentimes broader) should also be considered to avoid too broad a sweep of the technological field. In balancing these factors, the committee has identified a limited number of top technical conferences across four technology-specific research areas that, based on its expertise and deep domain knowledge of the field, are most critical to addressing the computing performance challenge described in Chapter 1.

*F.1.8 Methodological Summary*

In summary, the committee believes that high quality conferences attract leading researchers and showcase significant, recent research contributions to the field. In Sections F.2 and F.3, geographic distributions of conference authorship are quantified to provide a technology-specific assessment of national and regional research capabilities.[40,41]

The committee's analysis is not intended to be representative of all scientific outputs across the four specified research areas; for example, it does not presume that all relevant technologies presented in journal publications are implicitly represented in the selected conferences. The committee also recognizes the significant and ongoing progress in bibliometric and scientometric approaches to assess the quantity, quality, and impact of scientific output. The assessment provided in subsequent sections is provided as an important first step toward new approaches to assess the global research landscape in specific advanced computing technologies.

## F.2 Current National and Regional Advanced Research Capabilities in Four Key Technology Areas

In this section, conference representation is used as a proxy indicator of a nation's current (2011) advanced research capabilities in each of the following key technology areas: semiconductor devices and circuits, computer architecture, programming systems, and applications. For each technology area, all nations with at least 1 percent conference representation are ranked based on their weighted authorship contributions (following the methodology described in Section F.1.2) in the targeted conferences previously described in Sections F.1.3–F.1.6. In addition, regional research capabilities are also provided for comparison.

As previously discussed, conference papers tend to have a much shorter time from submission to publication than computing journals in related areas, resulting in the most recent, significant innovation appearing at conferences first. Thus, an assessment of current conference research efforts is particularly relevant given the increasing rate at which scientific discoveries are made and then disseminated via the Web. Time series analyses of national and regional research capabilities are shown in Section F.3.

*F.2.1 Advanced Semiconductor Devices and Circuit Research*

Table F-4 shows national capabilities in advanced semiconductor and nanoscale devices (IEDM and NANO) research, as well as semiconductor circuits (ISSCC) research in 2011. The table shows all countries with at least 1 percent representation. The United States has a strong competitive position in both of these areas (50 percent in devices and 36 percent in circuits) followed by Japan, Taiwan, and Korea. China has only token representation in these conferences at this time. Figures F-2 and F-3 show the same data broken down by region.

In semiconductor devices, the United States has the highest representation with half of the papers, followed by Asia with less than one-third of the papers and Europe with even fewer.[42] In circuits, the United States, Asia, and Europe all share approximately one third of the papers. NANO represents research that typically is targeted further in the future than those published in IEDM. At NANO the United States represents over 60

---

[40]This approach is consistent with a 2010 National Research Council (NRC) report, *S&T Strategies of Six Countries: Implications for the United States*, in which conference publication analysis was described as a "technology-specific indicator [that] gives a relatively accurate picture of [national] S&T standing." Available at http://www.nap.edu/catalog.php?record_id=12920.

[41]A related analysis was also conducted in a 2000 NRC report, *Experiments in International Benchmarking of U.S. Research Fields*, which used U.S. contributions of papers at the annual Conference on Magnetism and Magnetic Materials as a measure of U.S. participation in magnetic materials research. See http://www.nap.edu/catalog.php?record_id=9784.

[42]U.S. representation in semiconductor devices is reduced from ~50 percent to ~38 percent when only IEDM publications are considered (see Figure F-7 and Table F-8). Despite this reduction, the United States remains well ahead of Japan (~17 percent).

TABLE F-4 Current (2011) National Capabilities in Advanced Semiconductor and Nanoscale Devices and Circuits Research (Measured by Percent Share of Conference Papers)

| Advanced Semiconductor Circuits Research | | |
|---|---|---|
| 1 | USA | 36.4 |
| 2 | Japan | 12.4 |
| 3 | Korea | 10.4 |
| 4 | Netherlands | 7.3 |
| 5 | Taiwan | 6.7 |
| 6 | Germany | 5.3 |
| 7 | Belgium | 3.5 |
| 8 | France | 3.4 |
| 9 | Italy | 2.9 |
| 10 | UK | 2.3 |
| 11 | China | 2.0 |
| 12 | Canada | 1.8 |
| 13 | Switzerland | 1.6 |
| 14 | Austria | 1.0 |
| Advanced Semiconductor and Nanoscale Devices Research | | |
| 1 | USA | 50.2 |
| 2 | Japan | 10.4 |
| 3 | Taiwan | 6.8 |
| 4 | Korea | 5.7 |
| 5 | UK | 3.1 |
| 6 | Germany | 2.9 |
| 7 | Canada | 2.7 |
| 8 | France | 2.5 |
| 9 | Belgium | 2.2 |
| 10 | Italy | 2.1 |
| 11 | China | 1.9 |
| 12 | India | 1.8 |
| 13 | Singapore | 1.1 |
| 14 | UAE | 1.1 |

Data compiled from IEDM and NANO (semiconductor and nanoscale devices) and ISSCC (semiconductor circuits).

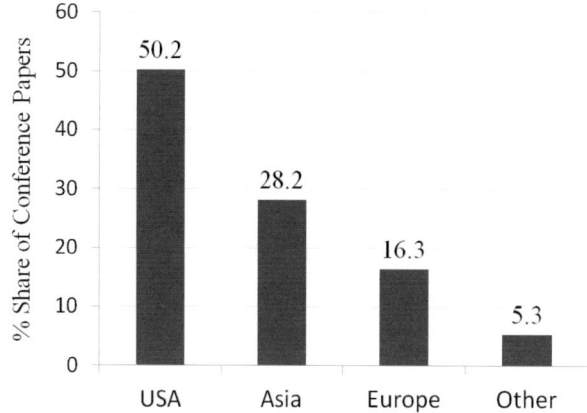

FIGURE F-2 Current (2011) regional capabilities in advanced semiconductor and nanoscale devices research (measured by percent share of conference papers). Data compiled from IEDM and NANO.

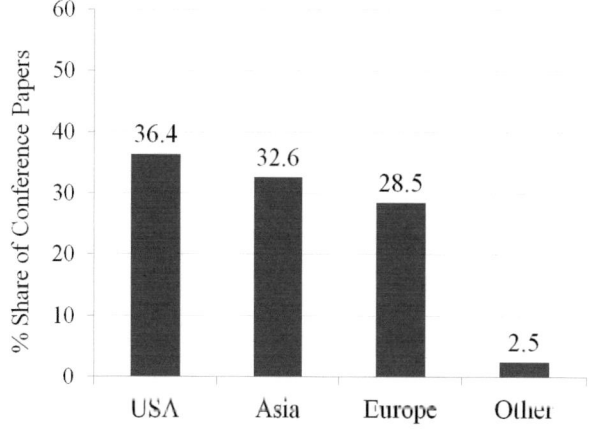

FIGURE F-3 Current (2011) regional capabilities in advanced semiconductor circuits research (measured by percent share of conference papers). Data compiled from ISSCC.

percent of all papers, while Asia and Europe represent less than 20 percent and 10 percent respectively.

*F.2.2 Advanced Architecture Research*

Table F-5 shows national and regional capabilities in advanced architecture research based on aggregated data from ASPLOS, HPCA, ISCA, and MICRO in 2011. This table includes all countries with at least 1 percent representation. Figure F-4 shows the same data broken down by region. U.S. research dominates that of any other nation or region and has a weighted average representation of more than 85 percent of the papers. No other nation or region contributes more than 3 percent or

7 percent of the papers, respectively. This result is perhaps not surprising, given the U.S. historical dominance in commercial microprocessors, including Intel, AMD, and IBM, as well as former commercial microprocessors from DEC, HP, and others. That Japan is not represented on this list suggests that Japanese universities and industry research institutions are not focused on mainstream computer architectures. While Japan has activity and expertise in the area, notably the custom processors from Fujitsu that are in the K supercomputer, their national research focus generally lies elsewhere.

TABLE F-5 Current (2011) National Capabilities in Advanced Architecture Research (Measured by Percent Share of Conference Papers)

| | Advanced Architecture Research | |
|---|---|---|
| 1 | USA | 85.7 |
| 2 | Korea | 2.5 |
| 3 | France | 2.2 |
| 4 | China | 1.7 |
| 5 | Canada | 1.6 |
| 6 | Switzerland | 1.4 |
| 7 | Australia | 1.0 |

Data compiled from ASPLOS, HPCA, ISCA, and MICRO.

### F.2.3 Advanced Programming Systems Research

Table F-6 shows national capabilities in advanced programming systems research based on aggregated data from ECOOP, OOPSLA, PLDI, POPL, and PPoPP. This table includes all countries with at least 1 percent representation. Figure F-5 shows the same data broken down by region. In programming systems research, the United States dominates with approximately 60 percent of the papers. The national breakdown in Table F-6 shows that aside from the United States, the national distribution of papers is diverse.

Figure F-5 shows that in programming systems, Europe is a distant second to the United States with about 25 percent representation, while all of Asia accounts for less than 10 percent. Disaggregating the data (see Appendix G), the United States has an even stronger position in PPoPP and PLDI, accounting for about 75 percent representation in each of them. These conferences are practical in nature, with papers presenting prototype software systems and applications running on real hardware platforms. In POPL, a more theoretical conference, the United States has a little less than 45 percent of the papers, on par with European presentation. Asia has less than 10 percent of the papers, with greater representation in applications than programming systems. In ECOOP, the United States accounted for about 30 percent representation, putting it 15 percentage points behind Europe.

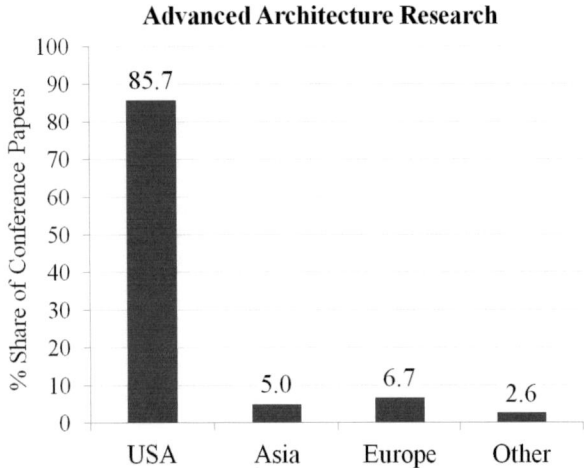

FIGURE F-4 Current (2011) regional capabilities in advanced architecture research (measured by percent share of conference papers). Data compiled from ASPLOS, HPCA, ISCA, and Micro.

TABLE F-6 Current (2011) National Capabilities in Advanced Programming Systems Research (Measured by Percent Share of Conference Papers)

| | Advanced Programming Systems Research | |
|---|---|---|
| 1 | USA | 36.4 |
| 2 | Germany | 12.4 |
| 3 | UK | 10.4 |
| 4 | Switzerland | 7.3 |
| 5 | Israel | 6.7 |
| 6 | France | 5.3 |
| 7 | Canada | 3.5 |
| 8 | India | 3.4 |
| 9 | Japan | 2.9 |
| 10 | China | 2.3 |
| 11 | Denmark | 2.0 |
| 12 | Chile | 1.8 |
| 13 | Korea | 1.6 |

Data compiled from ECOOP, OOPSLA, PLDI, POPL, and PPoPP.

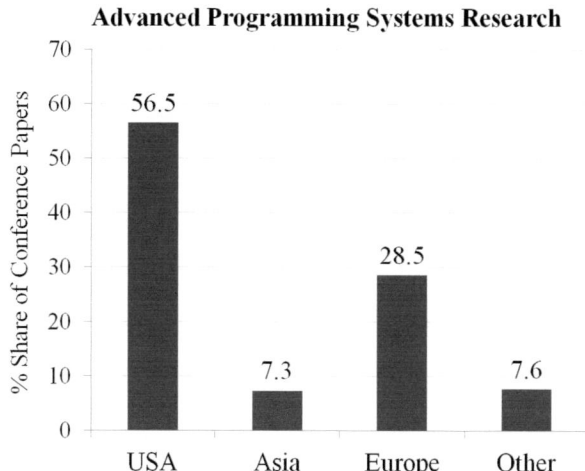

FIGURE F-5 Current (2011) regional capabilities in advanced programming systems research (measured by percent share of conference papers). Data compiled from ECOOP, OOPSLA, PLDI, POPL, and PPoPP.

*2.2.4 Advanced Applications Research*

Table F-7 shows national capabilities in advanced applications research based on aggregated data from Eurographics, OSDI, SIGGRAPH, SC, SOSP, VLDB, and WWW in 2011. Figure F-6 shows this data broken down into regions. In applications research, the U.S. accounted for almost 64 percent of papers, followed by Germany, Canada, and China, with 5-6 percent representation each.

Disaggregated data show that the United States is also the lead paper contributor in SOSP (~92 percent compared with ~4 percent in Asia), SC (~80 percent compared with 9 percent for both Asia and Europe), and WWW (~72 percent compared with ~13 percent in Europe and ~8 percent in Asia) and very strong leads in SIGGRAPH and VLDB (~53 percent, putting the United States ahead of Europe by approximately 30 points in both conferences), suggesting the United States maintains strong core competencies in parallel applications.

TABLE F-7 Current (2011) National Capabilities in Advanced Applications Research (Measured by Percent Share of Conference Papers)

| | Advanced Applications Research | |
|---|---|---|
| 1 | USA | 63.9 |
| 2 | Germany | 6.0 |
| 3 | Canada | 4.7 |
| 4 | China | 4.3 |
| 5 | France | 2.4 |
| 6 | Israel | 2.3 |
| 7 | UK | 2.2 |
| 8 | Hong Kong | 1.9 |
| 9 | Italy | 1.5 |
| 10 | Switzerland | 1.5 |
| 11 | Japan | 1.2 |
| 12 | Korea | 1.0 |

Data compiled from Eurographics, OSDI, SIGGRAPH, SC, SOSP, VLDB, and WWW.

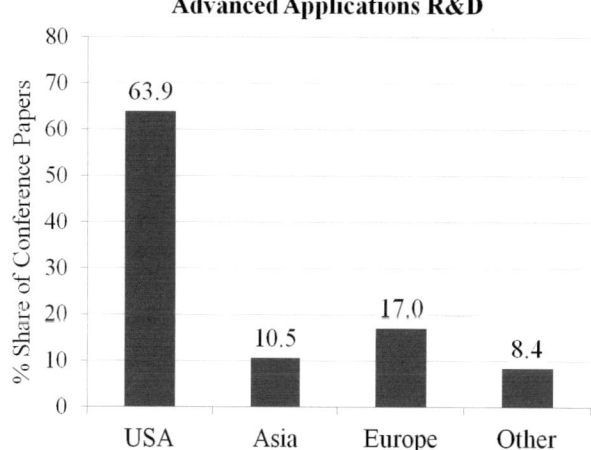

FIGURE F-6 Current (2011) regional capabilities in advanced applications research (measured by percent share of conference papers). Data compiled from Eurographics, OSDI, SIGGRAPH, SC, SOSP, VLDB, and WWW.

## F.2.5 Summary of Current National Technological Leadership

In each of the four key technical areas, the United States holds the lead, with particularly strong representation in architecture research ~85 percent (compared to ~5 and 7 percent representation by Asia and Europe, respectfully). In programming systems research, the United States maintains the lead at ~36 percent followed by Europe at ~29 percent and Asia at ~8 percent. The strongest European paper contributors were Germany and the UK at ~12 and 10 percent, respectfully. In applications research, the United States leads paper contributions at ~64 percent followed by Europe at ~17 percent and Asia at ~11 percent. This data suggests that the U.S. is particularly strong in design and systems engineering. While the United States maintains a strong lead in semiconductor and nanoscale devices (~50 percent) compared to Asia (~28 percent) and Europe (~16 percent), its contributions in semiconductor circuits research are comparable to Europe and Asia (each separated by ~7 percentage points).

## F.3 Longitudinal Changes in the Global Research Landscape

To assess how the competitive research landscape has changed over time, the committee examined the competitive metrics (described in Section F.1) for the same set of conferences in the previous section in 5-year increments over a 15-year span from 1996 to 2011.

Tables F-8 through F-11 show the relative representation of different nations in each of the technical areas of advanced computing research from 1996 to 2011 in 5-year increments, ranked by position in 2011. Each of the tables includes only those nations that have at least a 1 percent representation for at least one of the years. Countries that are not represented in 2011, but have at least 1 percent representation for any increment, are marked in italicized text. Figures F-7 through F-11 show the aggregate regional representation for each area of advanced research over the same time span. For the time-series data, the semiconductor devices area includes only IEDM, because historical data for NANO was not easily available.

In semiconductor devices research, U.S. representation has remained relatively stable with the largest gains made by Taiwan and Belgium (~ 5-6 percentage point increases each). In semiconductor circuits research, the United States shows a moderate decline, in tandem with an overall broadening in

TABLE F-8 (1996–2011) National Capabilities in Advanced Semiconductor and Devices and Circuits Research (Measured by Percent Share of Conference Papers)

| Semiconductors & Nanoscale Devices | | | | |
|---|---|---|---|---|
| | 1996 | 2001 | 2006 | 2011 |
| USA | 39.3 | 40.6 | 34.2 | 37.9 |
| Japan | 34.4 | 24.5 | 21.1 | 16.9 |
| Taiwan | 2.3 | 5.0 | 4.5 | 8.2 |
| Korea | 6.3 | 8.1 | 9.4 | 7.5 |
| Belgium | 1.5 | 3.0 | 6.4 | 6.4 |
| France | 3.4 | 2.3 | 4.4 | 6.0 |
| Italy | 3.9 | 3.4 | 2.8 | 3.7 |
| Singapore | - | 1.7 | 4.5 | 3.0 |
| China | 0.5 | 0.5 | 0.6 | 2.6 |
| UK | - | - | 1.7 | 1.9 |
| Germany | 3.0 | 5.4 | 3.7 | 1.7 |
| Austria | - | 0.4 | 0.6 | 1.1 |
| *Canada* | *0.5* | *1.0* | *0.4* | *0.1* |
| *Netherlands* | *1.9* | *1.7* | *1.7* | *0.7* |
| *Switzerland* | *2.1* | *0.8* | *2.2* | *0.6* |

| Semiconductor Circuits | | | | |
|---|---|---|---|---|
| | 1996 | 2001 | 2006 | 2011 |
| USA | 46.6 | 51.1 | 45.2 | 36.4 |
| Japan | 27.7 | 18.6 | 15.8 | 12.4 |
| Korea | 3.9 | 4.7 | 6.1 | 10.4 |
| Netherlands | 2.0 | 4.0 | 3.1 | 7.3 |
| Taiwan | 0.6 | 0.1 | 7.0 | 6.7 |
| Germany | 4.2 | 4.4 | 5.3 | 5.3 |
| Belgium | 1.0 | 5.5 | 1.5 | 3.5 |
| France | 0.9 | 0.8 | 1.0 | 3.4 |
| Italy | 2.9 | - | 3.4 | 2.9 |
| UK | 0.8 | 0.8 | 0.6 | 2.3 |
| China | - | - | 1.5 | 2.0 |
| Canada | 3.4 | 2.2 | 1.4 | 1.8 |
| Switzerland | 4.3 | 2.7 | 1.9 | 1.6 |
| Austria | - | - | 2.3 | 1.0 |
| *Finland* | *-* | *1.2* | *0.1* | *-* |
| *Hong Kong* | *-* | *1.2* | *-* | *-* |
| *Sweden* | *0.7* | *-* | *1.0* | *0.4* |

Data compiled from IEDM (semiconductor and nanoscale devices) and ISSCC (semiconductor circuits).

international representation. In this area largest leaps were made by Korea, Taiwan, and the Netherlands with 5–6 percentage point increases each. At the same time, Japan has dropped significantly in both semiconductor devices and in semiconductor circuits (~17 and 15 percent, respectfully).

In architecture research, the United States has maintained a significant lead, with no major advances by any other nation or region. In programming systems, the U.S. lead has been challenged somewhat by increases in Europe by small but steady gains by Israel, Switzerland, and the UK (as well as China, India, and Korea to a lesser degree).

In the application areas, U.S. representation has retained a stable lead over the 15-year period with no other nation ever contributing more than 8 percent (France, which contributed ~8 percent in 1996, has since dropped to ~2 percent in 2011). While representing only a small percentage of applications papers, China made a notable move from no representation in 1996 to ~4 percent in 2011.

TABLE F-9 (1996–2011) National Capabilities in Advanced Architecture Research (Measured by Percent Share of Conference Papers)

|  | 1996 | 2001 | 2006 | 2011 |
|---|---|---|---|---|
| USA | 79.9 | 89.2 | 90.7 | 85.7 |
| Korea | - | 0.2 | - | 2.5 |
| France | 3.1 | 1.0 | 0.7 | 2.2 |
| China | - | - | - | 1.7 |
| Canada | 5.5 | 1.2 | 1.9 | 1.6 |
| Switzerland | - | - | - | 1.4 |
| Australia | 0.4 | - | - | 1.0 |
| Belgium | - | 1.0 | 0.3 | - |
| India | - | - | 1.9 | 0.6 |
| Israel | - | - | 1.2 | - |
| Japan | 2.1 | 2.0 | 1.2 | - |
| Spain | 2.4 | 5.5 | 0.8 | 0.6 |
| Sweden | 1.8 | - | - | - |
| UK | 2.0 | - | 0.3 | 0.7 |

Data compiled from ASPLOS, HPCA, ISCA, and Micro.

TABLE F-10 (1996–2011) National Capabilities in Advanced Programming Systems Research (Measured by Percent Share of Conference Papers)

|  | 1996 | 2001 | 2006 | 2011 |
|---|---|---|---|---|
| USA | 62.2 | 63.8 | 67.1 | 56.5 |
| Germany | 8.0 | 3.0 | 2.3 | 7.9 |
| UK | 5.2 | 4.3 | 7.1 | 7.2 |
| Switzerland | 1.9 | 2.9 | 1.8 | 4.5 |
| Israel | 1.7 | 5.1 | 2.0 | 3.2 |
| France | 7.5 | 4.8 | 3.1 | 3.0 |
| India | - | - | - | 2.1 |
| Japan | 2.7 | 5.2 | 2.3 | 2.0 |
| Canada | 2.4 | 1.0 | 1.7 | 2.2 |
| China | - | - | 1.2 | 1.4 |
| Denmark | 2.4 | 2.3 | 0.9 | 1.4 |
| Chile | - | - | 0.2 | 1.1 |
| Korea | - | - | 0.4 | 1.0 |
| Australia | 0.4 | 0.6 | 1.8 | 0.7 |
| Netherlands | 2.3 | 2.1 | 0.8 | 0.9 |

Data compiled from ECOOP, OOPSLA, PLDI, POPL, and PPoPP.

TABLE F-11 (1996–2011) National Capabilities in Advanced Applications Systems Research (Measured by Percent Share of Conference Papers)

|  | 1996 | 2001 | 2006 | 2011 |
|---|---|---|---|---|
| USA | 64.3 | 57.5 | 64.0 | 63.9 |
| Germany | 4.2 | 7.3 | 7.9 | 6.0 |
| Canada | 3.5 | 3.7 | 3.5 | 4.7 |
| China | - | 0.5 | 3.0 | 4.3 |
| France | 8.3 | 3.5 | 1.9 | 2.4 |
| Israel | 1.5 | 2.1 | 1.4 | 2.3 |
| UK | 1.8 | 3.0 | 3.9 | 2.2 |
| Hong Kong | - | 0.3 | 1.7 | 1.9 |
| Italy | 1.8 | 1.7 | 1.3 | 1.5 |
| Switzerland | 1.4 | 1.9 | 0.7 | 1.5 |
| Japan | 3.7 | 5.6 | 3.0 | 1.2 |
| Korea | 0.4 | 1.1 | 0.7 | 1.0 |
| Australia | 0.7 | 1.0 | 0.4 | 0.4 |
| Austria | 1.3 | 1.5 | 0.5 | 0.7 |
| Singapore | 0.3 | 1.4 | 0.6 | 0.9 |
| Taiwan | 1.0 | 0.9 | 0.8 | 0.6 |

Data compiled from Eurographics, OSDI, SIGGRAPH, SC, SOSP, VLDB, and WWW.

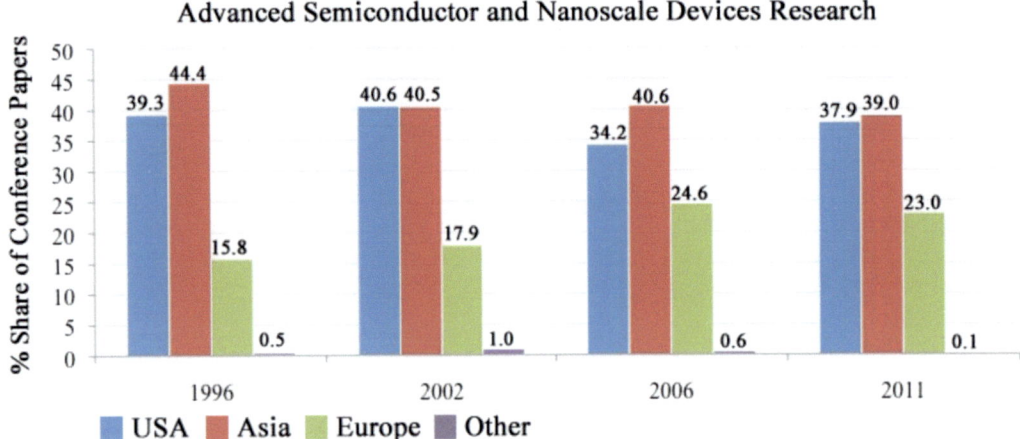

FIGURE F-7 (1996–2011) Regional capabilities in advanced semiconductor and nanoscale devices research (measured by percent share of conference papers). Data compiled from IEDM.

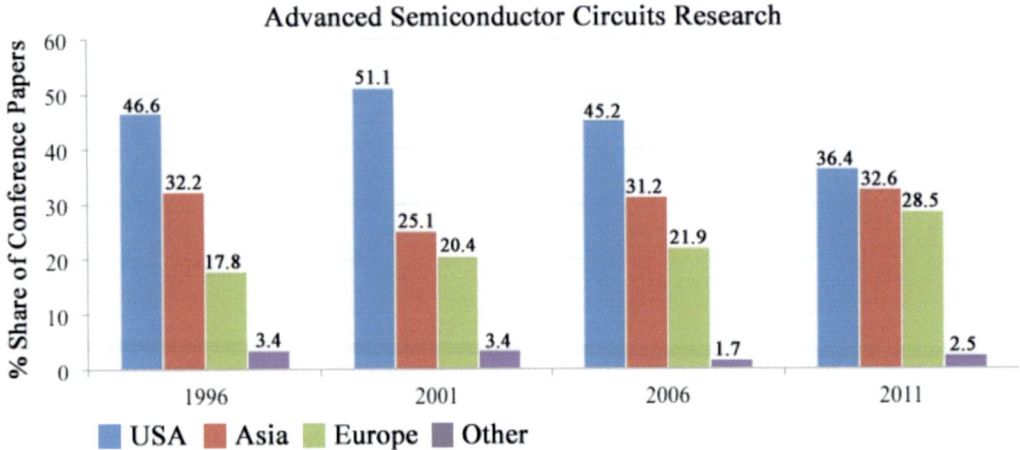

FIGURE F-8 (1996–2011) Regional capabilities in advanced semiconductor circuits research (measured by percent share of conference papers). Data compiled from ISSCC.

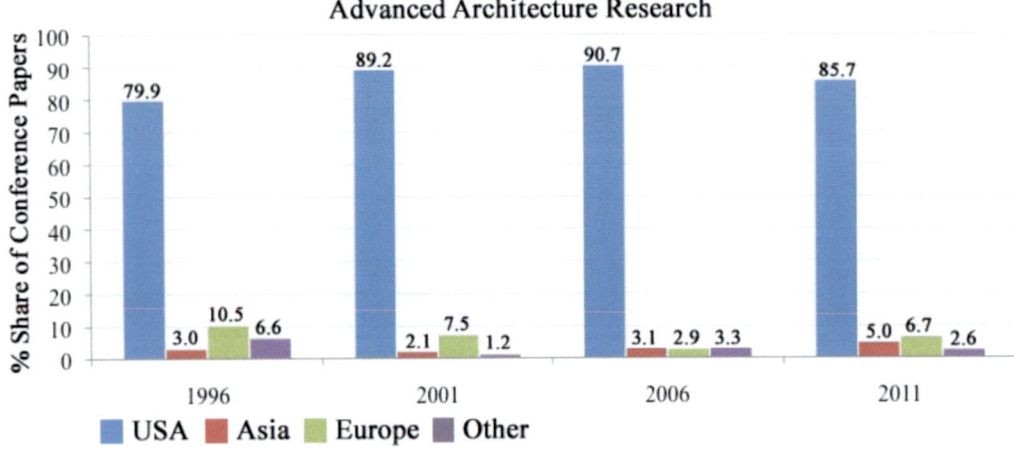

FIGURE F-9 (1996–2011) Regional capabilities in advanced architecture research (measured by percent share of conference papers). Data compiled from ASPLOS, HPCA, ISCA, and MICRO.

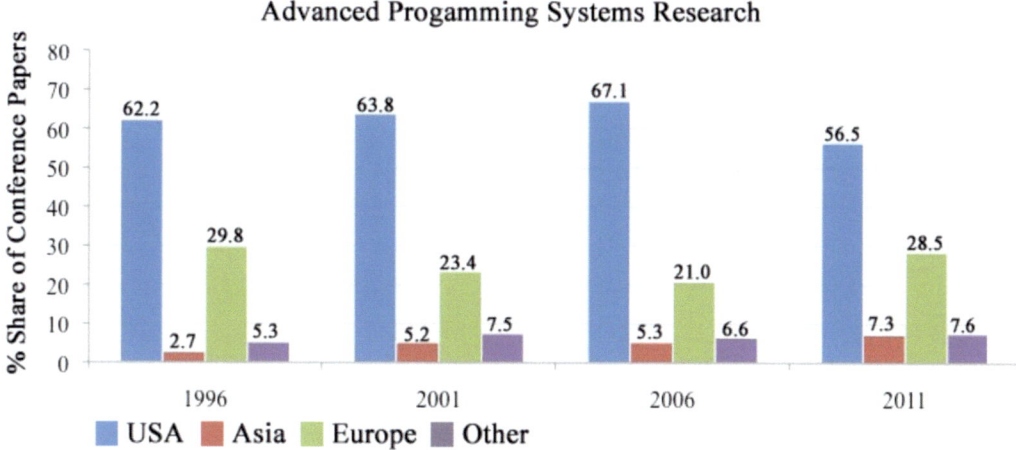

FIGURE F-10 (1996–2011) Regional capabilities in advanced programming systems research (measured by percent share of conference papers). Data compiled from ECOOP, OOPSLA, PLDI, POPL, and PPoPP.

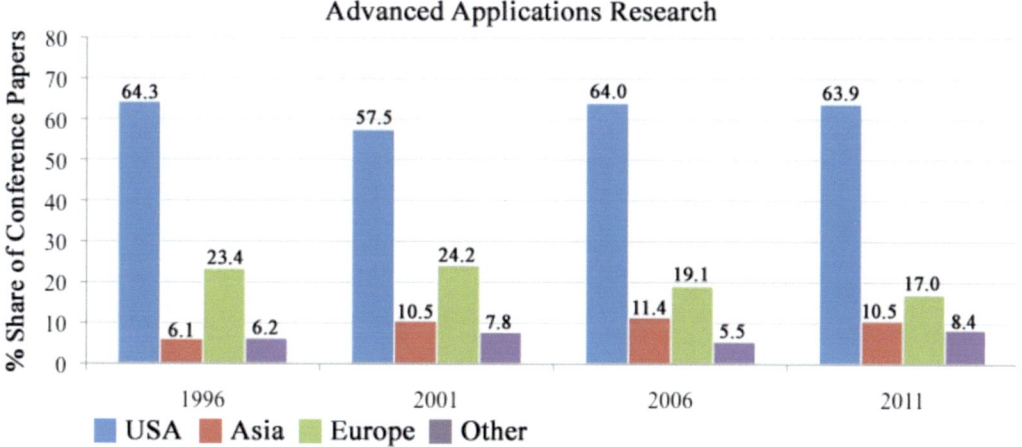

FIGURE F-11 (1996–2011) Regional capabilities in advanced applications research (measured by percent share of conference papers). Data compiled from Eurographics, OSDI, SC, SIGGRAPH, SOSP, VLDB, and WWW.

# G

# Conference Bibliometric Data

TABLE G-1 Conference Statistics

|  | Papers | | | | Authors | | | |
|---:|---:|---:|---:|---:|---:|---:|---:|---:|
|  | 1996 | 2001 | 2006 | 2011 | 1996 | 2001 | 2006 | 2011 |
| ASPLOS | 25 | 24[#] | 41 | 32 | 99 | 82 | 169 | 152 |
| ECOOP | 21 | 18 | 25 | 26 | 47 | 48 | 64 | 86 |
| Eurographics | 43 | 52 | 44 | 35 | 112 | 145 | 151 | 140 |
| HPCA | 29 | 26 | 29 | 46 | 79 | 89 | 107 | 182 |
| ISCA | 28 | 23 | 31 | 40 | 94 | 74 | 124 | 181 |
| IEDM | 206 | 220[#] | 226 | 219 | 1135 | 1921 | 2391 | 2020 |
| ISSCC | 155 | 165 | 253 | 211 | 1078 | 1237 | 1691 | 1622 |
| MICRO | 30 | 29 | 42 | 44 | 81 | 98 | 181 | 177 |
| NANO | - | - | - | 409 | - | - | - | 1488 |
| OOPSLA | 26 | 27 | 26 | 61 | 75 | 78 | 96 | 216 |
| OSDI | 19 | 22[*] | 27 | 32[Δ] | 62 | 83 | 127 | 152 |
| PLDI | 28 | 30 | 36 | 55 | 73 | 80 | 127 | 216 |
| POPL | 34 | 26 | 36 | 49 | 67 | 45 | 93 | 149 |
| PPoPP | 25[+] | 14 | 26 | 26 | 68 | 40 | 97 | 104 |
| SC | 54 | 58 | 54 | 74 | 180 | 208 | 273 | 370 |
| SIGGRAPH | 58 | 64 | 87 | 115 | 163 | 213 | 322 | 447 |
| SOSP | 23[+] | 17 | 25[±] | 28 | 99 | 58 | 123 | 154 |
| VLDB | 48 | 59 | 78 | 104 | 141 | 109 | 266 | 386 |
| WWW | 58 | 78 | 84 | 81 | 136 | 247 | 276 | 295 |

4,719 Papers Analyzed                23,859 Authors Analyzed

*Conferences analyzed in adjacent years are marked as follows: [+] held in 1997; [*] held in 2000; [#] held in 2002; [+] held in 2007; and [Δ] held in 2010*

TABLE G-2 Advanced Semiconductor and Nanoscale Devices and Circuits Conferences (Weighted Percent Share of Papers)

| ISSCC held in: | 1996 USA | 2001 USA | 2006 USA | 2011 USA | IEDM held in: | 1996 USA | 2002 USA | 2006 USA | 2011 USA | NANO held in: | 2011 USA |
|---|---|---|---|---|---|---|---|---|---|---|---|
| USA | 46.6 | 51.1 | 45.2 | 36.4 | USA | 39.3 | 40.6 | 34.2 | 37.9 | USA | 56.8 |
| China | - | - | 1.5 | 2.0 | China | 0.5 | 0.5 | 0.6 | 2.6 | China | 1.5 |
| Hong Kong | - | 1.2 | - | - | Hong Kong | 0.6 | 0.5 | - | 0.8 | Hong Kong | 0.1 |
| India | - | - | - | 0.9 | India | 0.2 | 0.2 | 0.6 | - | India | 2.7 |
| Japan | 27.7 | 18.6 | 15.8 | 12.4 | Japan | 34.4 | 24.5 | 21.1 | 16.9 | Japan | 6.9 |
| Korea | 3.9 | 4.7 | 6.1 | 10.4 | Korea | 6.3 | 8.1 | 9.4 | 7.5 | Korea | 4.7 |
| Singapore | - | 0.5 | 0.8 | 0.2 | Singapore | - | 1.7 | 4.5 | 3.0 | Malaysia | 0.2 |
| Taiwan | 0.6 | 0.1 | 7.0 | 6.7 | Taiwan | 2.4 | 5.0 | 4.5 | 8.2 | Taiwan | 6.1 |
| Austria | - | - | 2.3 | 1.0 | Austria | - | 0.4 | 0.6 | 1.1 | France | 0.6 |
| Belgium | 1.0 | 5.5 | 1.5 | 3.5 | Belgium | 1.5 | 3.0 | 6.4 | 6.4 | Germany | 3.6 |
| Denmark | - | 0.6 | 0.6 | 0.1 | Finland | - | 0.1 | - | - | Italy | 1.2 |
| Finland | - | 1.2 | 0.1 | - | France | 3.4 | 2.3 | 4.4 | 6.0 | Netherlands | 0.6 |
| France | 0.9 | 0.8 | 1.0 | 3.4 | Germany | 3.0 | 5.4 | 3.7 | 1.7 | Norway | 1.0 |
| Germany | 4.2 | 4.4 | 5.3 | 5.3 | Ireland | - | 0.5 | - | - | Poland | 0.9 |
| Greece | - | 0.1 | 0.4 | - | Italy | 3.9 | 3.4 | 2.8 | 3.7 | Russia | 0.5 |
| Ireland | 0.6 | - | - | 0.2 | Liechtenstein | - | 0.1 | - | - | Spain | 0.5 |
| Italy | 2.9 | - | 3.4 | 2.9 | Netherlands | 1.9 | 1.7 | 1.7 | 0.7 | Switzerland | 0.2 |
| Netherlands | 2.0 | 4.0 | 3.1 | 7.3 | Poland | - | - | 0.1 | - | Turkey | 0.1 |
| Norway | - | 0.1 | - | 0.2 | Spain | - | 0.4 | 0.5 | 0.5 | UK | 3.7 |
| Portugal | - | - | 0.7 | 0.3 | Sweden | - | - | 0.5 | 0.5 | Iran | 0.2 |
| Spain | 0.3 | - | - | - | Switzerland | 2.1 | 0.8 | 2.2 | 0.6 | Lebanon | 0.2 |
| Sweden | 0.7 | - | 1.0 | 0.4 | UK | - | - | 1.7 | 1.9 | ▓▓▓▓ | 0.1 |
| Switzerland | 4.3 | 2.7 | 1.9 | 1.6 | Israel | - | - | .1 | - | UAE | 1.6 |
| Turkey | - | 0.1 | - | - | Canada | 0.5 | 1.0 | 0.4 | 0.1 | S. Africa | 0.5 |
| UK | 0.8 | 0.8 | 0.6 | 2.3 | | | | | | Australia | 0.7 |
| Israel | | 0.6 | 0.2 | 0.6 | Regional Distribution | | | | | Brazil | 0.2 |
| Canada | 3.4 | 2.2 | 1.4 | 1.8 | USA | 39.3 | 40.6 | 34.2 | 37.9 | Colombia | 0.2 |
| Australia | - | 0.6 | - | - | Asia | 44.4 | 40.5 | 40.6 | 39.0 | Canada | 4.2 |
| | | | | | Europe | 15.8 | 17.9 | 24.6 | 23.0 | | |
| Regional Distribution | | | | | Other | 0.5 | 1.0 | 0.6 | 0.1 | Regional Distribution | |
| USA | 46.6 | 51.1 | 45.2 | 36.4 | | | | | | USA | 56.8 |
| Asia | 32.2 | 25.1 | 31.2 | 32.6 | | | | | | Asia | 22.4 |
| Europe | 17.8 | 20.4 | 21.9 | 28.5 | | | | | | Europe | 12.8 |
| Other | 3.4 | 3.4 | 1.7 | 2.5 | | | | | | Other | 8.1 |

TABLE G-3 Advanced Architecture Conferences (Weighted Percent Share of Papers)

| ASPLOS | 1996 | 2002 | 2006 | 2011 | HPCA | 1996 | 2001 | 2006 | 2011 |
|---|---|---|---|---|---|---|---|---|---|
| held in: | USA | USA | USA | USA | held in: | USA | Mexico | USA | USA |
| USA | 79.5 | 95.1 | 90.7 | 83.3 | USA | 67.4 | 83.7 | 89.5 | 88.5 |
| Japan | 1.6 | 4.2 | 2.4 | - | China | - | - | - | 2.4 |
| Korea | - | - | - | 1.6 | Hong Kong | 3.4 | - | - | - |
| Belgium | - | - | 1.2 | - | India | - | - | 5.2 | 1.4 |
| Denmark | - | - | - | 0.5 | Japan | 3.4 | - | - | - |
| France | 6.0 | - | - | 2.6 | Belgium | - | 3.8 | - | - |
| Germany | - | - | 2.4 | 3.1 | France | - | 3.8 | - | 2.7 |
| Norway | 0.8 | - | - | - | Germany | 0.7 | - | - | - |
| Spain | - | 0.7 | - | - | Greece | 3.4 | - | - | - |
| Switzerland | - | - | - | 3.1 | Spain | 3.4 | 7.7 | 2.8 | 1.3 |
| UK | - | - | 0.3 | - | Sweden | 6.9 | - | - | - |
| Australia | - | - | - | 5.0 | Switzerland | - | - | - | 1.4 |
| Brazil | 3.3 | - | - | - | UK | 4.1 | - | - | - |
| Canada | 8.8 | - | 2.4 | 0.8 | Australia | 1.4 | - | - | - |
| Jordan | - | - | 0.5 | - | Canada | 5.7 | 1.0 | - | 2.2 |
|  |  |  |  |  | Israel | - | - | 2.6 | - |
|  | Regional Distribution | | | | | | | | |
| USA | 79.5 | 95.1 | 90.7 | 83.3 |  | Regional Distribution | | | |
| Asia | 1.6 | 4.2 | 2.4 | 1.6 | USA | 67.4 | 83.7 | 89.5 | 88.5 |
| Europe | 6.8 | 0.7 | 4.0 | 9.4 | Asia | 6.9 | - | 5.2 | 3.9 |
| Other | 12.1 | - | 2.9 | 5.8 | Europe | 18.6 | 15.4 | 2.8 | 5.4 |
|  |  |  |  |  | Other | 7.1 | 1.0 | 2.6 | 2.2 |

| MICRO | 1996 | 2001 | 2006 | 2011 | ISCA | 1996 | 2001 | 2006 | 2011 |
|---|---|---|---|---|---|---|---|---|---|
| held in: | France | USA | USA | Brazil | held in: | USA | Sweden | USA | USA |
| USA | 90.0 | 86.2 | 88.3 | 83.0 | USA | 82.4 | 93.1 | 94.9 | 87.6 |
| India | - | - | 2.9 | - | China | - | - | - | 4.2 |
| Japan | - | 3.4 | - | - | India | - | - | - | 0.8 |
| Korea | - | - | - | 5.3 | Japan | 3.6 | - | 2.4 | - |
| Singapore | - | - | - | 0.6 | Korea | - | 0.7 | - | 2.9 |
| France | - | - | 2.4 | 2.3 | France | 7.1 | - | - | 1.3 |
| Greece | - | - | - | 1.4 | Italy | - | - | - | 2.5 |
| Spain | 3.3 | 6.9 | 0.8 | 0.9 | Spain | 2.4 | 6.2 | - | - |
| Switzerland | - | - | - | 1.5 | UK | - | - | 1.2 | - |
| UK | 3.3 | - | - | 2.7 | Canada | 4.5 | - | 1.5 | 0.8 |
| Canada | 3.3 | 3.4 | 3.2 | 2.3 |  |  |  |  |  |
| Israel | - | - | 2.4 | - |  | Regional Distribution | | | |
|  |  |  |  |  | USA | 82.4 | 93.1 | 94.9 | 87.6 |
|  | Regional Distribution | | | | Asia | 3.6 | 0.7 | 2.4 | 7.9 |
| USA | 90.0 | 86.2 | 88.3 | 83.0 | Europe | 9.5 | 6.2 | 1.2 | 3.8 |
| Asia | - | 3.4 | 2.9 | 6.0 | Other | 4.5 | - | 1.5 | 0.8 |
| Europe | 6.7 | 6.9 | 3.2 | 8.8 |  |  |  |  |  |
| Other | 3.3 | 3.4 | 5.6 | 2.3 |  |  |  |  |  |

TABLE G-4 Advanced Programming Systems Conferences (Weighted Percent Share of Papers)

| ECOOP held in: | 1996 Austria | 2001 Hungary | 2006 France | 2011 UK |
|---|---|---|---|---|
| USA | 26.2 | 38.0 | 56.3 | 30.3 |
| China | - | - | - | 3.1 |
| India | - | - | - | 3.8 |
| Japan | 4.8 | 5.6 | 4.0 | - |
| Belgium | - | 3.7 | 8.0 | - |
| Denmark | 7.1 | 5.6 | 4.0 | 3.8 |
| France | 14.3 | 16.7 | - | - |
| Germany | 23.8 | - | - | 19.2 |
| Ireland | - | - | 1.0 | 2.2 |
| Italy | - | 7.4 | - | - |
| Netherlands | 6.3 | 3.7 | - | - |
| Norway | - | 5.6 | - | - |
| Sweden | 3.2 | - | - | - |
| Switzerland | 7.1 | - | 3.0 | 13.5 |
| UK | - | - | 13.0 | 7.7 |
| Australia | 2.4 | 1.9 | 4.0 | - |
| Canada | - | 0.9 | 1.3 | 5.1 |
| Chile | - | - | - | 8.7 |
| Israel | 4.8 | 5.6 | - | - |
| New Zealand | - | 5.6 | 5.3 | 2.6 |
| Regional Distribution | | | | |
| USA | 26.2 | 38.0 | 56.3 | 30.3 |
| Asia | 4.8 | 5.6 | 4.0 | 6.9 |
| Europe | 61.9 | 42.6 | 29.0 | 46.4 |
| Other | 7.1 | 13.9 | 10.7 | 16.3 |

| OOPSLA held in: | 1996 USA | 2001 USA | 2006 USA | 2011 USA |
|---|---|---|---|---|
| USA | 56.4 | 58.6 | 52.1 | 53.1 |
| China | - | - | - | 0.8 |
| Hong Kong | - | - | 3.8 | - |
| India | - | - | - | 2.0 |
| Japan | 3.8 | 11.1 | 3.8 | 3.3 |
| Korea | - | - | - | 1.1 |
| Singapore | - | - | 3.8 | 1.6 |
| Austria | - | - | - | 0.7 |
| Belgium | - | - | 3.8 | - |
| Czech Republic | 3.8 | - | - | - |
| Denmark | - | 1.2 | - | 0.9 |
| Estonia | - | - | - | 0.4 |
| France | 11.5 | - | - | 0.8 |
| Germany | 7.7 | 7.4 | 8.5 | 7.4 |
| Italy | - | 3.7 | - | 1.6 |
| Netherlands | 6.7 | - | 3.8 | 3.3 |
| Portugal | - | - | - | 0.3 |
| Romania | 1.0 | - | - | - |
| Switzerland | - | 8.6 | - | 9.9 |
| UK | 3.8 | - | 14.7 | 4.7 |
| Australia | - | - | 0.2 | 1.0 |
| Brazil | 1.3 | - | - | - |
| Canada | - | 3.7 | - | 1.6 |
| Chile | - | - | 1.3 | 0.3 |
| Israel | 3.8 | 5.6 | 3.8 | 5.2 |
| Regional Distribution | | | | |
| USA | 56.4 | 58.6 | 52.1 | 53.1 |
| Asia | 3.8 | 11.1 | 11.5 | 8.8 |
| Europe | 34.6 | 21.0 | 31.0 | 30.0 |
| Other | 5.1 | 9.3 | 5.4 | 8.1 |

| PLDI held in: | 1996 USA | 2001 USA | 2006 Canada | 2011 USA |
|---|---|---|---|---|
| USA | 86.3 | 93.3 | 79.0 | 75.1 |
| China | - | - | 0.6 | 1.2 |
| India | - | - | - | 3.9 |
| Japan | - | - | 4.2 | 3.6 |
| Singapore | - | - | - | 1.8 |
| Austria | - | 3.3 | 0.9 | - |
| Denmark | - | - | - | - |
| France | - | - | - | 2.2 |
| Germany | 5.4 | - | 2.4 | 1.8 |
| Ireland | - | - | 1.9 | - |
| Italy | 1.8 | - | - | 1.8 |
| Spain | - | - | 0.6 | - |
| UK | 3.6 | - | 2.0 | 4.9 |
| Australia | - | - | 1.9 | - |
| Canada | 3.0 | - | 1.1 | 1.5 |
| Israel | - | 3.3 | 5.6 | 2.0 |
| Regional Distribution | | | | |
| USA | 86.3 | 93.3 | 79.0 | 75.1 |
| Asia | - | - | 4.7 | 10.6 |
| Europe | 12.3 | 3.3 | 7.8 | 10.7 |
| Other | 3.0 | 3.3 | 8.5 | 3.5 |

TABLE G-4 Advanced Programming Systems Conferences (continued)

| POPL | 1996 | 2001 | 2006 | 2011 | PPoPP | 1997 | 2001 | 2006 | 2011 |
|---|---|---|---|---|---|---|---|---|---|
| held in: | USA | UK | USA | USA | held in: | USA | USA | USA | USA |
| USA | 54.4 | 44.2 | 58.3 | 43.5 | USA | 82.1 | 80.0 | 88.3 | 75.9 |
| China | - | - | 2.4 | - | China | - | - | 2.6 | 3.8 |
| Japan | 2.0 | 3.8 | - | 0.5 | India | - | - | - | 0.8 |
| Korea | - | - | 1.9 | 1.0 | Japan | 4.0 | 7.1 | - | - |
| Austria | - | - | - | 1.0 | Korea | - | - | - | 3.8 |
| Belgium | - | - | - | 2.0 | France | 8.0 | - | 1.0 | - |
| Denmark | 4.9 | 1.3 | 0.9 | 3.1 | Germany | 1.6 | - | - | 3.8 |
| France | 5.9 | 9.6 | 12.0 | 9.9 | Greece | - | - | 1.0 | 3.8 |
| Germany | 5.4 | 5.8 | 0.9 | 11.6 | Netherlands | - | 12.9 | - | - |
| Italy | 2.9 | 1.9 | - | - | Portugal | - | - | - | 3.8 |
| Netherlands | - | - | 0.5 | - | Spain | - | - | 0.5 | - |
| Sweden | - | - | 6.9 | 1.0 | Brazil | 2.7 | - | - | - |
| Switzerland | 2.9 | 3.8 | 5.6 | 0.3 | Canada | 1.6 | - | 6.7 | - |
| Turkey | - | - | - | 0.5 | Israel | - | - | - | 3.8 |
| UK | 14.7 | 19.2 | 7.9 | 16.6 | Saudi Arabia | - | - | - | 0.3 |
| Israel | 1.0 | 9.0 | - | 3.4 | | | | | |
| Australia | - | 1.3 | 2.8 | 2.0 | Regional Distribution | | | | |
| Canada | 5.9 | - | - | 3.4 | USA | 82.1 | 80.0 | 88.3 | 75.9 |
| | | | | | Asia | 4.0 | 7.1 | 2.6 | 8.5 |
| Regional Distribution | | | | | Europe | 9.6 | 12.9 | 2.4 | 11.5 |
| USA | 54.4 | 44.2 | 58.3 | 43.5 | Other | 4.3 | - | 6.7 | 4.1 |
| Asia | 2.0 | 3.8 | 4.2 | 1.5 | | | | | |
| Europe | 36.8 | 41.7 | 34.7 | 46.1 | | | | | |
| Other | 6.9 | 10.3 | 2.8 | 8.8 | | | | | |

TABLE G-5 Advanced Applications Conferences (Weighted Percent Share of Papers)

| Eurographics held in: | 1996 UK | 2001 UK | 2006 Austria | 2011 UK |
|---|---|---|---|---|
| USA | 12.4 | 14.9 | 23.2 | 32.3 |
| China | - | - | 4.5 | 7.6 |
| Hong Kong | - | - | - | 2.3 |
| Japan | 4.7 | 15.4 | 6.8 | 1.0 |
| Korea | - | 1.9 | 3.8 | 2.9 |
| Taiwan | 2.3 | 1.9 | 2.3 | - |
| Austria | 9.3 | 5.8 | - | 2.1 |
| Denmark | 2.3 | - | - | - |
| Finland | 2.3 | - | 2.3 | - |
| France | 32.6 | 9.6 | 10.2 | 6.3 |
| Germany | 8.1 | 18.9 | 28.0 | 24.0 |
| Hungary | - | 1.9 | - | 2.9 |
| Italy | - | 1.9 | - | - |
| Netherlands | 1.7 | - | - | - |
| Poland | - | - | - | 0.6 |
| Russia | - | 3.2 | - | - |
| Spain | 2.3 | 1.4 | - | 2.4 |
| Sweden | - | - | - | 1.0 |
| Switzerland | 2.3 | 8.7 | - | 1.9 |
| UK | 4.7 | 3.8 | 8.7 | 1.9 |
| Australia | - | - | 0.8 | - |
| Brazil | - | - | 0.5 | - |
| Canada | 4.7 | 2.9 | 7.6 | 8.1 |
| Israel | 7.9 | 5.8 | 1.4 | 2.9 |
| S. Africa | 2.3 | 1.9 | - | - |
| Regional Distribution | | | | |
| USA | 12.4 | 14.9 | 23.2 | 32.3 |
| Asia | 7.0 | 19.2 | 17.4 | 13.7 |
| Europe | 65.7 | 55.3 | 49.2 | 43.0 |
| Other | 14.9 | 10.6 | 10.2 | 11.0 |

| OSDI held in: | 1996 USA | 2000 USA | 2006 USA | 2010 Canada |
|---|---|---|---|---|
| USA | 89.5 | 86.4 | 94.8 | 86.8 |
| Japan | - | - | - | 0.9 |
| Korea | - | 4.5 | - | - |
| France | - | 4.5 | - | - |
| Germany | - | - | - | 2.8 |
| Portugal | 5.3 | - | 0.7 | - |
| Spain | - | 4.5 | - | - |
| UK | - | - | - | 0.8 |
| Australia | - | - | - | 3.1 |
| Canada | 5.3 | - | 3.7 | 3.1 |
| Israel | - | - | 0.7 | 2.4 |
| Regional Distribution | | | | |
| USA | 89.5 | 86.4 | 94.8 | 86.8 |
| Asia | - | 4.5 | - | 0.9 |
| Europe | 5.3 | 9.1 | 0.7 | 3.6 |
| Other | 5.3 | - | 4.4 | 8.7 |

| SC held in: | 1996 USA | 2001 USA | 2006 USA | 2011 USA |
|---|---|---|---|---|
| USA | 84.4 | 73.1 | 88.7 | 80.4 |
| China | - | - | 1.9 | 2.3 |
| India | - | - | 1.9 | 0.3 |
| Japan | 6.2 | 7.2 | 1.4 | 4.7 |
| Korea | 2.0 | - | - | 0.6 |
| Taiwan | - | 1.7 | - | 1.0 |
| Austria | - | 2.9 | - | 1.9 |
| Belgium | - | - | - | 1.4 |
| Czech Republic | - | 0.6 | - | - |
| France | - | - | 1.9 | 3.3 |
| Germany | 1.2 | 2.5 | 1.9 | 0.7 |
| Ireland | - | - | - | 0.2 |
| Italy | 1.9 | 1.7 | - | - |
| Netherlands | - | - | - | 0.5 |
| Spain | 3.7 | 4.0 | 1.9 | 0.5 |
| Sweden | - | 1.7 | - | - |
| Switzerland | 0.6 | - | - | 0.8 |
| UK | - | 3.3 | 0.6 | - |
| Australia | - | 1.3 | - | - |
| Saudi Arabia | - | - | - | 1.4 |
| Regional Distribution | | | | |
| USA | 84.4 | 73.1 | 88.7 | 80.4 |
| Asia | 8.2 | 8.9 | 5.1 | 9.0 |
| Europe | 7.4 | 16.7 | 6.2 | 9.2 |
| Other | - | 1.3 | - | 1.4 |

TABLE G-5 Advanced Applications Conferences (continued)

| **SIGGRAPH** | 1996 | 2001 | 2006 | 2011 | **SOSP** | 1997 | 2001 | 2007 | 2011 |
|---|---|---|---|---|---|---|---|---|---|
| held in: | USA | USA | USA | Canada | held in: | France | Canada | USA | Portugal |
| USA | 83.6 | 63.8 | 66.8 | 52.7 | USA | 96.5 | 94.4 | 87.7 | 92.2 |
| China | - | 1.6 | 5.8 | 5.0 | China | - | - | - | 4.2 |
| Hong Kong | - | - | 2.3 | 2.1 | Germany | - | - | 9.1 | - |
| Japan | 5.2 | 7.8 | 3.6 | 1.3 | Switzerland | 3.5 | - | 3.2 | 0.9 |
| Korea | - | - | 1.1 | 1.6 | UK | | 5.6 | | |
| Singapore | - | - | - | 0.8 | Canada | - | - | - | 2.0 |
| Taiwan | 1.7 | - | 1.1 | 0.9 | Israel | - | - | - | 0.7 |
| Austria | - | - | 0.9 | - | | | | | |
| Belgium | - | - | 0.6 | 0.4 | | Regional Distribution | | | |
| Finland | - | - | 1.1 | - | USA | 96.5 | 94.4 | 87.7 | 92.2 |
| France | - | 3.6 | 1.1 | 4.7 | Asia | - | - | - | 4.2 |
| Germany | 0.9 | 8.6 | 7.2 | 10.0 | Europe | 3.5 | 5.6 | 12.3 | 0.9 |
| Greece | - | - | - | 0.1 | Other | - | - | - | 2.8 |
| Spain | - | - | - | 0.2 | | | | | |
| Sweden | - | - | 1.1 | 1.6 | | | | | |
| Switzerland | 1.7 | - | 1.1 | 2.0 | | | | | |
| UK | - | 0.9 | 1.8 | 4.1 | | | | | |
| Australia | - | 0.7 | - | - | | | | | |
| Brazil | - | 0.5 | - | 1.1 | | | | | |
| Canada | 6.9 | 12.5 | 2.0 | 5.5 | | | | | |
| Israel | - | - | 2.1 | 5.3 | | | | | |
| Saudi Arabia | - | - | - | 0.5 | | | | | |
| | Regional Distribution | | | | | | | | |
| USA | 83.6 | 63.8 | 66.8 | 52.7 | | | | | |
| Asia | 6.9 | 9.4 | 14.0 | 11.8 | | | | | |
| Europe | 2.6 | 13.1 | 15.1 | 23.2 | | | | | |
| Other | 6.9 | 13.7 | 4.1 | 12.4 | | | | | |

TABLE G-5 Advanced Applications Conferences (continued)

| VLDB held in: | 1996 India | 2001 Italy | 2006 Korea | 2011 USA | WWW held in: | 1996 France | 2001 China | 2006 UK | 2011 India |
|---|---|---|---|---|---|---|---|---|---|
| USA | 57.6 | 54.4 | 62.8 | 53.8 | USA | 49.1 | 54.0 | 50.9 | 72.5 |
| China | - | - | - | 4.3 | China | - | 0.9 | 4.7 | 5.4 |
| Hong Kong | - | 1.7 | 3.8 | 5.4 | Hong Kong | - | - | 2.0 | 0.2 |
| India | 4.2 | 1.7 | 2.8 | 0.9 | India | - | 2.3 | 4.5 | 2.2 |
| Japan | 4.2 | 3.4 | 1.3 | - | Japan | 1.7 | 0.4 | 4.8 | - |
| Korea | - | 3.4 | - | 1.5 | Singapore | - | 1.3 | 0.4 | 0.2 |
| Taiwan | - | - | 1.3 | 0.7 | Taiwan | 1.7 | 1.3 | - | - |
| Singapore | 2.1 | 6.4 | 2.6 | 2.9 | Austria | - | - | 1.2 | 1.4 |
| Austria | - | 1.1 | - | - | Belgium | - | - | 1.2 | - |
| Belgium | - | - | 0.6 | - | Czech Repub | 1.7 | - | 1.2 | - |
| Denmark | 0.7 | - | - | 1.8 | Denmark | - | 1.3 | - | - |
| Finland | 2.1 | - | - | - | France | 12.1 | 0.6 | - | - |
| France | 8.3 | 5.9 | 1.3 | 1.0 | Germany | 7.8 | 4.5 | 5.9 | 2.2 |
| Germany | 7.3 | 8.8 | 5.9 | 4.8 | Greece | 3.4 | 1.3 | - | 1.0 |
| Greece | - | 1.7 | 1.3 | 1.3 | Hungary | - | - | 1.2 | - |
| Italy | 4.2 | 1.7 | 2.6 | 3.4 | Ireland | - | - | 1.2 | - |
| Netherlands | 2.1 | - | 0.3 | - | Italy | 4.3 | 3.8 | 3.9 | 4.5 |
| Norway | - | 1.7 | - | 1.0 | Netherlands | 1.7 | 4.1 | 0.3 | - |
| Sweden | - | - | - | 1.0 | Portugal | - | - | 0.2 | - |
| Switzerland | 2.1 | 1.7 | 1.3 | 2.0 | Russia | - | 1.3 | - | - |
| UK | - | 1.7 | 2.9 | 3.1 | Spain | - | 1.3 | 1.2 | 0.2 |
| Australia | 2.1 | - | 0.3 | 0.7 | Sweden | 1.7 | 2.1 | - | - |
| Brazil | - | - | - | 1.0 | Switzerland | - | 1.3 | - | 1.6 |
| Canada | 3.1 | 3.1 | 7.7 | 7.5 | UK | 6.0 | 6.3 | 8.9 | 1.5 |
| Israel | - | 1.7 | 1.3 | 1.6 | Australia | 1.7 | 3.0 | 1.2 | - |
| Qatar | - | - | - | 0.2 | Argentina | - | 0.4 | - | - |
| | | | | | Brazil | - | 1.1 | 0.5 | - |
| | Regional Distribution | | | | Canada | 3.4 | 2.0 | 2.4 | 4.5 |
| USA | 57.6 | 54.4 | 62.8 | 53.8 | Chile | - | - | - | 0.8 |
| Asia | 10.4 | 16.5 | 11.8 | 15.8 | S. Africa | - | 1.3 | - | - |
| Europe | 26.7 | 24.3 | 16.1 | 19.3 | Israel | 1.7 | 4.2 | 2.4 | 1.5 |
| Other | 5.2 | 4.7 | 9.3 | 11.0 | New Zealand | 1.7 | - | - | - |
| | | | | | | Regional Distribution | | | |
| | | | | | USA | 49.1 | 54.0 | 50.9 | 72.5 |
| | | | | | Asia | 3.4 | 6.2 | 16.3 | 8.0 |
| | | | | | Europe | 38.8 | 27.8 | 26.4 | 12.6 |
| | | | | | Other | 8.6 | 12.0 | 6.4 | 6.9 |

# H

# Top 20 Largest Hardware and Software Companies

Table H-1 World's 20 Largest Hardware and Software Companies in 2010 (in U.S. $ Millions)

| Rank | Hardware Companies (Country) | Hardware Revenue | Total Revenue | Rank | Software Companies (Country) | Software Revenue | Total Revenues |
|---|---|---|---|---|---|---|---|
| 1 | Samsung (S. Korea) | 77,865 | 120,119 | 1 | Microsoft (USA) | 49,090 | 61,159 |
| 2 | HP (USA) | 73,729 | 116,245 | 2 | IBM (USA) | 21,396 | 95,758 |
| 3 | Foxconn (Taiwan) | 44,411 | 44,411 | 3 | Oracle (USA) | 18,582 | 22,734 |
| 4 | LG Electronics (S Korea) | 42,029 | 63,043 | 4 | SAP (Germany) | 11,386 | 15,373 |
| 5 | Nokia (Finland) | 40,108 | 59,042 | 5 | Ericsson (Sweden) | 7,595 | 29,014 |
| 6 | Toshiba (Japan) | 40,057 | 69,778 | 6 | Nintendo (Japan) | 6,799 | 17,726 |
| 7 | Dell (USA) | 38,395 | 53,585 | 7 | HP (USA) | 6,183 | 116,245 |
| 8 | Intel (USA) | 34,026 | 35,172 | 8 | Symantec (USA) | 5,565 | 5,992 |
| 9 | Apple (USA) | 31,772 | 43,086 | 9 | Nokia Siemens Networks (Finland) | 4,529 | 18,114 |
| 10 | Cisco (USA) | 29,510 | 36,633 | 10 | Activision Blizzard (USA) | 4,279 | 4,279 |
| 11 | Quanta Computer (Taiwan) | 24,755 | 24,755 | 11 | CA (USA) | 4,012 | 4,318 |
| 12 | Fujitsu (Japan) | 23,056 | 50,662 | 12 | EMC (USA) | 3,960 | 14,026 |
| 13 | Canon (Japan) | 22,567 | 34,719 | 13 | Electronic Arts (USA) | 3,728 | 3,728 |
| 14 | Ricoh (Japan) | 19,484 | 22,018 | 14 | Adobe (USA) | 2,796 | 2,987 |
| 15 | Asus (Taiwan) | 19,074 | 19,074 | 15 | Cisco (USA) | 2,137 | 36,663 |
| 16 | Acer (Taiwan) | 17,944 | 17,944 | 16 | SunGard (USA) | 1,996 | 5,508 |
| 17 | Compal Electronics (Taiwan) | 16,923 | 19,909 | 17 | Sony (Japan) | 1,914 | 79,441 |
| 18 | IBM (USA) | 16,190 | 95,758 | 18 | BMC (USA) | 1,758 | 1,888 |
| 19 | Lenovo (China) | 16,132 | 16,132 | 19 | Alcatel-Lucent (USA) | 1,635 | 21,835 |
| 20 | NEC (Japan) | 16,127 | 40,475 | 20 | Konami (Japan) | 1,594 | 2,887 |

*Adopted from www.hardwaretop100.org*     *Adopted from www.softwaretop100.org*

*The methodology employed in creating these tables is available at http://www.hardwaretop100.org/methodology.php. Last accessed on June 16, 2012. While many of the companies on the list are global in nature, the table lists each company's nation of origin.*

# I

# China's Medium- and Long-Term Plan

Table I-1 Key Areas, Technologies, and Programs Identified in China's Medium- and Long-Term Plan for Development of Science and Technology

| Key Areas (11): | Frontier Technologies (8): |
|---|---|
| <ul><li>Agriculture</li><li>Energy</li><li>Environment</li><li>Information technology industry and modern services</li><li>Manufacturing</li><li>National defense</li><li>Population and health</li><li>Public securities</li><li>Transportation</li><li>Urbanization and urban development</li><li>Water and mineral resources</li></ul> | <ul><li>Advanced energy</li><li>Advanced manufacturing</li><li>Aerospace and aeronautics</li><li>Biotechnology</li><li>Information</li><li>Laser</li><li>New materials</li><li>Ocean</li></ul> |
| **Engineering Megaprojects*  (16):** | **Science Megaprojects (4):** |
| <ul><li>Advanced numeric-controlled machinery and basic manufacturing technology</li><li>Control and treatment of AIDS, hepatitis, and other major diseases</li><li>Core electronic components, high-end generic chips, and basic software</li><li>Drug innovation and development</li><li>Extra- large-scale integrated circuit manufacturing and technique</li><li>Genetically modified new-organism variety breeding</li><li>High-definition Earth observation systems</li><li>Large advanced nuclear reactors</li><li>Large aircraft</li><li>Large-scale oil and gas exploration</li><li>Manned aerospace and Moon exploration</li><li>New-generation broadband wireless mobile telecommunications</li><li>Water pollution control and treatment</li></ul> | <ul><li>Development and reproductive biology</li><li>Nanotechnology</li><li>Protein science</li><li>Quantum research</li></ul> |

*The MLP only marks 13 megaengineering programs. The rest are presumably military-related programs.

Adapted from C. Cao et al., 2006, China's 15-year science and technology plan, *Physics Today*, p. 38.

# J

# List of Abbreviations

| | |
|---|---|
| AC | Advanced Computing |
| ACM | Association for Computing Machinery |
| ACSAC | Asia-Pacific Computer Systems Architecture Conference |
| ASPLOS | International Symposium on Architectural Support for Programming Languages and Operating Systems |
| A-SSCC | Asian Solid-State Circuits Conference |
| ATP | Advanced Technology Products |
| CAGR | Compound Annual Growth Rate |
| CAS | Chinese Academy of Sciences |
| CCMA | Cloud Computing Center for Mobile Application |
| CMOS | Complementary-Symmetry Metal-Oxide Semiconductors |
| Code 4 | Information and Communications Advanced Technology Products |
| Code 5 | Electronics Advanced Technology Products |
| COTS | Commercial-Off-the-Shelf |
| CPU | Central Processing Unit |
| DARPA | Defense Advanced Research Projects Agency |
| DOD | Department of Defense |
| EC | European Commission |
| ECFA | Economic Cooperation Framework Agreement |
| ECOOP | European Conference on Object-Oriented Programming |
| ESSCIRC | European Solid-State Circuits Conference |
| ESSDERC | European Solid-State Device Research Conference |
| Eurographics | Conference of the European Association for Computer Graphics |
| FGCS | Fifth Generation Computer Systems (project) |
| FP7 | Seventh Framework Program (2007–2013) |
| FPGA | Field-Programmable Gate Arrays |
| GDP | Gross Domestic Product |
| GPU | Graphics Processing Unit |
| GP-GPU | General Purpose Graphics Processing Unit |
| HiPEAC | European Network of Excellence on High-Performance and Embedded Architectures and Compilers |
| HPC | High-Performance Computing |
| HPCA | International Symposium on High Performance Computer Architecture |
| IC | Integrated Circuits |
| ICT | Information and Communications Technology |
| IDM | Integrated Device Manufacturer |

| | |
|---|---|
| IEDM | International Electron Devices Meeting |
| IEEE | Institute of Electrical and Electronics Engineers |
| ILP | Instruction-Level Parallelism |
| IP | Internet Protocol |
| IPC | Instructions per Clock Cycle |
| IPR | Intellectual Property Rights |
| ISA | Instruction Set Architecture |
| ISC | International Symposium on Computer Architecture |
| ISSCC | International Solid-State Circuits Conference |
| IT | Information Technology |
| ITRI | Industrial Technology Research Institute |
| KET | Key Enabling Technologies |
| LED | Light-emitting Diode |
| MCC | Microelectronics and Computer Technology Consortium |
| MEMS | Microelectromechanical Systems |
| MICRO | International Symposium on Microarchitecture |
| MIIT | Ministry of Industry and Information Technology |
| MLP | Medium- and Long-Term Plan (2006–2020) |
| MOST | Ministry of Science and Technology |
| NANO | International Conference on Nanotechnology |
| NNIN | National Nanofabrication Infrastructure Network |
| NRC | National Research Council |
| NSF | National Science Foundation |
| ODM | Original Device Manufacturer |
| OEM | Original Equipment Manufacturer |
| OOPSLA | Object-Oriented Programming, Systems, Languages, and Applications |
| O-S-D | Optoelectronics-Sensor-Discrete |
| OSDI | Symposium on Operating Systems Design and Implementation |
| PARLE | Parallel Architectures and Languages Europe |
| PC | Personal Computer |
| PLDI | Programming Language Design and Implementation |
| POPL | Symposium on Principles of Programming Languages |
| PPoPP | Symposium on Principles and Practice of Parallel Programming |
| PwC | PricewaterhouseCoopers |
| R&D | Research and Development |
| S&T | Science and Technology |
| SC | International Conference for High Performance Computing, Networking, Storage, and Analysis |
| $Sci^2$ | Science of Science Tool |
| SEI | Strategic Emerging Industries |
| SEMATECH | Semiconductor Manufacturing Technology |
| SIA | Semiconductor Industry Association |
| SIGGRAPH | International Conference on Computer Graphics and Interactive Techniques |
| SoC | System-on-a-Chip |
| SOSP | Symposium on Operating Systems Principles |
| SPA&T | Semiconductor Packaging, Assembly, and Test |
| SPERC | Secure, Parallel, Evolvable, Reliable and Correct (software) |
| SRC | Semiconductor Research Corporation |
| TSMC | Taiwan Semiconductor Manufacturing Company |
| VLDB | International Conference on Very Large Databases |
| WWW | International World Wide Web Conference |